内蒙古师范大学基本科研业务费专项资金（批准号：2022JBHQ015）资助
国家自然科学基金（批准号：62161045）资助
内蒙古自治区自然科学基金（批准号：2023LHMS01007）资助

Sturm-Liouville
Problems and Their Applications

施图姆-刘维尔问题及其应用

王桂霞　玉　林　著

图书在版编目(CIP)数据

施图姆-刘维尔问题及其应用 = Sturm-Liouville Problems and Their Applications / 王桂霞,玉林著. -- 厦门:厦门大学出版社,2023.11
ISBN 978-7-5615-9186-4

Ⅰ.①施… Ⅱ.①王… ②玉… Ⅲ.①常微分方程-数值计算 Ⅳ.①O241.81

中国版本图书馆CIP数据核字(2023)第222124号

出版人	郑文礼
责任编辑	眭 蔚
美术编辑	李嘉彬
技术编辑	许克华

出版发行	厦门大学出版社
社 址	厦门市软件园二期望海路39号
邮政编码	361008
总 机	0592-2181111 0592-2181406(传真)
营销中心	0592-2184458 0592-2181365
网 址	http://www.xmupress.com
邮 箱	xmup@xmupress.com
印 刷	厦门市明亮彩印有限公司

开本 720 mm×1 020 mm 1/16
印张 9
字数 242千字
版次 2023年11月第1版
印次 2023年11月第1次印刷
定价 48.00元

本书如有印装质量问题请直接寄承印厂调换

厦门大学出版社
微信二维码

厦门大学出版社
微博二维码

Preface

In 1836-1837, Sturm and Liouville created the Sturm-Liouville theory. In 1910, Weyl started to research singular Sturm-Liouville problems. In the 1950s, Shen Youcheng introduced the Sturm-Liouville theory to China. It was not until the late 1970s that Cao Zhijiang and Liu Jinglin organized a seminar at Inner Mongolia University, and the theory of ordinary differential operators began to be systematically studied in China. Up to now, many research results have been achieved. *Ordinary Differential Operators*[1], *Sturm-Liouville Problems and Its Inverse Problems*[2] and *The Geometric Aspects of Sturm-Liouville Problems*[3] were selected as textbooks or reference books for the students in China.

In recent years, discontinuous Sturm-Liouville problems with dependent eigenparameters boundary conditions have attracted the attention of many experts and scholars. More and more applications of the discontinuous Sturm-Liouville problems require quantitative information about the eigenvalues and eigenfunctions. For the eigenvalues, one is interested in both the values and the indices of the eigenvalues. But there are relatively few references related to numerical methods and the theory of odd-order differential operators. This book provides both theoretical results and algorithm implementations[4-6].

This book is available for senior undergraduate students and postgraduate students majoring in applied mathematics and computational mathematics. It can also serve as a reference for professionals in relevant fields.

At this time, we would like to acknowledge some people who have helped us with various aspects of this book. We thank Inner Mongolia Normal University and Inner Mongolia University for providing the scientific research platforms. We also thank the many students for their participation in editing and proofreading. This book is dedicated to the authors's tutors.

Though all of us tried, there are surely still some types and errors of thought. We take the blame for all of these mistakes. We would appreciate it if you would send any mistakes that you find to wgx@imnu.edu.cn or yulin@imnu.edu.cn.

This work is supported by the National Nature Science Foundation of China (62161045), the Fundamental Research Funds for the Inner Mongolia Normal University, China (2022JBHQ015), and the Natural Science Foundation of Inner Mongolia Autonomous Region, China (2023LHMS01007).

Wang Guixia
Mar., 2023

Contents

Chapter 1 Introduction 1
  1.1 Physical background 1
  1.2 Related results of ordinary differential operators 3
  1.3 Structure of the book 5

Chapter 2 Approximations of eigenvalues and eigenfunctions ... 10
  2.1 Notation and theoretic results 10
  2.2 Main ideas of the algorithms 16
  2.3 General methods for constructing examples 17
  2.4 Examples with λ-independent BCs 21
  2.5 Examples with λ-dependent BCs 27
  2.6 Oscillations of eigenfunctions for discontinuous Sturm-Liouville problems 30

Chapter 3 Computing the indices of eigenvalues 39
  3.1 Notation and theoretic results 39
  3.2 Algorithm and implementation 45
  3.3 Examples with a positive f 48
  3.4 Examples with an indefinite f 55
  3.5 Examples about $\beta_*(\alpha)$ and $\beta(\alpha)$ ... 61

Chapter 4 Relations among eigenvalues of Sturm-Liouville problems 67
  4.1 Notation and basic results 68
  4.2 Geometric characterization of λ_n 69
  4.3 Interlacing relations among eigenvalues 74

Chapter 5 Third-order eigenparameter dependent differential operators 78
  5.1 Preliminaries 78

5.2 The Banach space 80

5.3 Derivative formulas of eigenvalues 82

Chapter 6   Application of Sturm-Liouville problems **89**

6.1 Construction and stability of Riesz bases 89

6.2 Eigenvalue problems of internal solitary waves 98

Appendix A   Fundamentals Sturm-Liouville problems **106**

A.1 Classes of Sturm-Liouville problems 106

A.2 Characteristic function 112

Appendix B   Thomson-Haskell method **115**

Appendix C   First-order linear differential equations **119**

C.1 Existence and uniqueness of a solution 119

C.2 Rank of a solution and variation of parameters 124

C.3 Continuous dependence of solution on the problem 127

References **130**

Chapter 1 Introduction

The ordinary differential operator is an important class of unbounded linear operators. Its extensive usefulness could be found in various scenarios, such as beam vibration problems in engineering structures, the vertical structure of ocean internal solitary waves, or drug-eluting stent technology in medicine[7-11]. The theory of ordinary differential operators is a systematic and extensive mathematical branch integrating diverse theories and methods such as ordinary differential equations, functional analysis, and operator algebra, etc. The discipline of ordinary differential operators includes deficiency index theory, self-adjoint extension, qualitative and quantitative analysis of spectrum, numerical methods, completeness of eigenfunction systems, dependence of eigenvalues, asymptotic estimation, inverse spectral problems, and so on.

1.1 Physical background

The research of Sturm-Liouville (S-L) operators can be traced back to the 19th century, when Fourier established a mathematical model to solve the problem of heat conduction. In 1836-1837, Sturm and Liouville obtained a series of academic achievements on the spectral problems of regular second-order differential operators, which laid the foundation of the S-L theory. Nowadays, the classical S-L theory has become more and more complete, acting as an essential theoretical part in many applied sciences. As is known, the boundary conditions of the classical S-L operators are eigenparameters independent, and the functions acted by the operator in the maximal operator domain is at least continuous, however, these conditions can't be satisfied in some practical mathematical models.

For example, the string vibration problem with damping, the drug diffusion problem, and the heat conduction problem of thin plates formed by overlapping different materials[12-17]. Mathematical models of the above problems are discontinuous differential operator problems with transmission conditions. In recent years, research on differential operators with internal discontinuity has aroused the interest of many experts and scholars[18-27]. The internal discontinuity means that the solution to the differential equation or its quasi-derivative (or high-order quasi-derivative) is discontinuous at some points in the definition interval (i.e. discontinuous points).

Further research shows that many practical physical problems are governed by the S-L problem with eigenparameter boundary conditions. These practical physical problems include the wave propagation problem when sound waves meet different obstacles in underwater propagation, the string vibration problem caused by the motion of a simple pendulum, in which the string is fixed on the ceiling at one end, a spring is tied at the other end, and an object with a certain mass is fixed under the spring, etc.

As we all know, seismic waves propagating inside the earth are called body waves. Body wave includes P-wave and S-wave. The boundary between the mantle and different materials in the upper and lower layers is called a discontinuity. When P-wave and S-wave encounter discontinuous surfaces, the velocity will change obviously. Let u denote the displacement of the particle, $u' = du/dr$, $u(R_m^+) = \lim_{r \to R_m+0} u(r)$, $u(R_m^-) = \lim_{r \to R_m-0} u(r)$. ω denotes the characteristic frequency, l denotes the angular order, R_c is the radius of the core, R_m is the distance from the center of the earth to the Moho discontinuity, R denotes the radius of the earth, and $R_c < R_m < R$. Given interval J, the density of the earth $\rho(r)$ and S-wave velocity $\beta(r)$ are twice continuously differentiable in every compact subset of J. By substituting $U = ru$ in the equation for the torsional modes for a spherically symmetric without considering the rotation, Hald O. obtained the eigenvalue problem

$$-(r^4\rho\beta^2 u')' + (\ell+2)(\ell-1)r^2\rho\beta^2 u = \omega^2 r^4 \rho u, \ r \in I,$$
$$u'(R_c) = u'(R) = 0, \quad (1.1.1)$$

with

$$u(R_m^+) = u(R_m^-), \ (r^4\rho\beta^2 u')(R_m^+) = (r^4\rho\beta^2 u')(R_m^-)$$

at $r = R_m$. Using the Liouville transformation[28], transform (1.1.1) into the eigenvalue problems consisting of

$$-y'' + qy = \omega^2 K^2 y, \ x \in (0,\pi), x \in (0,d) \cup (d,\pi),$$

boundary conditions

$$y'(0) - hy(0) = y'(\pi) + Hy'(\pi) = 0,$$

and transmission conditions

$$y(d^+) = ay(d^-), \ y'(d^+) = a^{-1}y'(d^-) + by(d^-),$$

where

$$q(x) = f''(x)/f(x) + K^2(\ell+2)(\ell-1)\beta^2/r^2, \ f(x) = r^2(\rho(r)\beta(r))^{1/2},$$

$$K = \frac{1}{\pi} \int_{R_c}^{R} \frac{1}{\beta(r)} \mathrm{d}r, \ 0 < d < \frac{1}{2}\pi, \ |a-1| + |b| > 0.$$

1.2 Related results of ordinary differential operators

In 1954, Coddington gave an analytic description of the self-adjoint domain of higher-order symmetric differential operators in finite closed intervals, which is a complete description [29]. According to the general construction principle of self-adjoint operators, Naimark gave the necessary and sufficient conditions for the boundary condition coefficients of self-adjoint domains of symmetric differential operators defined by quasi-derivatives in finite closed intervals [30]. The description of self-adjoint domains of singular differential operators is essentially different from that of regular differential operators. In 1910, Weyl started the investigation of singular S-L problems, and proved that the second-order singular S-L problems can be classified into the limit point type and the limit circle type. Using the circular nesting method proposed by Weyl, Titchmarsh introduced the Weyl-Titchmarsh domain [31,32], for the self-adjoint domain of high-order singular symmetric differential operators of limit point type and limit circle type. It was not until the 1960s and 1970s, Everitt gave its analytical description using the matrix method [20,21,33]. In 1985, Cao gave a complete description of the self-adjoint extension of second-order singular differential operators, and presented a direct and complete description of the self-adjoint domain of high-order limit circle differential operators [1,34,35]. In 1986, Sun gave complete characterization of the self-adjoint domain of the differential operator with the middle deficiency indices [36]. In 2002, Wang studied the classification of the self-adjoint domain of differential operators for the first time, and gave the symplectic structure of the self-adjoint domain of differential operators [37]. In 2004, Wei obtained a necessary and sufficient condition for singular S-L differential operators to be a bound-limited self-adjoint extension, and characterized all forms of bound-limited self-adjoint extensions by complete classification of the self-adjoint boundary conditions [38]. [39,40] gave the standard forms of the self-adjoint boundary conditions for fourth-order regular and singular differential operators, respectively.

The research on the dependence of eigenvalues and eigenfunctions is of great significance in the differential operator theory, which provides the theoretical foundation for numerical computation of eigenvalues [41–43]. In 1993, Dauge proved that the eigenvalues of the regular S-L problem with Neumann boundary conditions are differentiable functions to the endpoints of the interval [44]. In 1996, Kong demonstrated the continuous differentiability (with respect to the boundary condition) of certain continuous eigenvalue branches and provided formulae for their differentials in several cases [45,46]. In [47], it was proved that the eigenvalues of linear regular two-point boundary value problems depend continuously on the problem, and the derivatives of the eigenvalues with respect to the given parameter were obtained. In [48], Kong proved that the n-th eigenvalue depends continuously on the problem, characterized the discontinuity set of λ_n as a function on the boundary conditions and determined the behavior of λ_n near each discontinuity point, and discussed the differentiability or analyticity of λ_n. In general, λ_n is also not differentiable at a self-adjoint BC where it is continuous and has multiplicity 2. [49] consider some geometric aspects of regular S-L problems, and investigate the relationships between the algebraic and geometric multiplicities of an eigenvalue. Since then, the dependence of eigenvalues has attracted many scholars, including higher-order differential operators with transmission conditions, S-L problems with eigenparameters dependent boundary conditions, and S-L problems on time scales [22,50-53], etc.

Most of the above work is about even-order differential operators. Such problems are well understood for even-order differential operators, especially for the so-called S-L problems [30,54-70]. However, little is known about odd order differential operators. Third-order differential equations arise in many physical phenomena, for example, the three-layer beam problem [8,9], more backgrounds can refer to [71]. In 2017, Hao characterized the self-adjoint domain of odd-order differential operators by using real parameter solutions [72], and Niu gave all canonical forms of self-adjoint BCs for regular third-order differential operators [73]. [14, 74–77] also studied a class of third-order self-adjoint boundary value problems and introduced the dependence of eigenvalues on the problem, and generalized these results to boundary value problems with discontinuity.

For problems of second-order S-L operators, refer to [78–82], when the discontinuous S-L problem is studied, it is usually regarded as the corresponding differential operator problem in the direct sum space. Everett described the self-adjoint domain of differential operators in direct sum spaces, and extended the problem of two interval direct sum spaces to the case of finite direct sum spaces and countable infinite direct sum spaces [20,21]. Sun generalized the conclusion in Chapter 13 of the book [69], and defined a new inner product in combination with the transmission condition, and established a new Hilbert space with appropriate parameters. In this space, they characterized all the self-adjoint realizations for the singular S-L problems in two intervals, in which the determinant of the coefficient matrix of the real coupling boundary condition is any positive number [26]. Mukhtarov extended the spectral properties of the regular S-L problem to a special discontinuous boundary value problem with boundary conditions dependent on spectral parameters and two supplementary transmission conditions [24]. Zhao further extended the above conclusions to the case of infinite discontinuous points and gave a characterization of self-adjoint fields[83]. In addition, there are also a lot of other research results on S-L problems with internal discontinuities, see [20, 50, 84–89].

The rapid development of computer technology has had a profound impact on mathematical research. SLEDGE[90], SLEIGN[41] SLEIGN2[91] are all highly effective codes for computing eigenvalues of regular scalar S-L problems with separated, self-adjoint boundary conditions. These problems may be either regular or singular. An algorithm in [92] is presented for computing the eigenvalues of regular self-adjoint S-L problems with matrix coefficients and arbitrary coupled boundary conditions. The mathematica codes in [6, 93] are used not only to approximate the eigenvalues for coupled conditions but also to determine their indices. The problem can be either self-adjoint, non-self-adjoint, or discontinuous; while the boundary conditions of the problem can depend on the spectral parameter. It's worth mentioning that inequalities among the eigenvalues of S-L problems with an arbitrary coupled boundary condition and two separated boundary conditions[94] play an important role in the mathematica codes above. [95] applied the sinc method, based on the sampling theory to approximate the eigenvalues of discontinuous S-L problems with the eigenparameter-dependent boundary conditions. Through corresponding numerical experiments, the theoretical research of spectral analysis has been promoted in turn. By analyzing the numerical examples, we can directly "see" the structural form of the eigenfunctions firstly, and subsequently conside and prove the basic properties such as the oscillation characterization of the eigenfunctions, which provides a research idea for the structural properties of the differential operator[27,96].

1.3 Structure of the book

This book contains six chapters and three appendices.

Many numerical methods have been developed for approximating eigenvalues and eigenfunctions of S-L problems. These techniques include the important finite difference, variational approaches, shooting methods, the utilization of Prüfer transformation and inequalities among eigenvalues, etc. However, the purpose of these methods is to approximate the eigenvalue when its index is given, specifically for self-adjoint S-L problems with λ-independent boundary conditions. Given an S-L problem, which can be either self-adjoint or non-self-adjoint; while the boundary conditions of the problem can depend on the spectral parameter, how to approximate the eigenvalues of the problem on a chosen interval, say $[-100,100]$, or a given rectangle region, say $[-20,50]\times[-40,80]$ efficiently?

Chapter 2 presents an idea for approximating the eigenvalues inside any given rectangle region: divide the given rectangle region into small rectangle regions, throw away the small regions that do not contain any eigenvalues and subdivide each remaining small region further; when this region-dividing-procedure is applied enough times, approximations of all eigenvalues of the problem in the given region can be found. We would like to emphasize that these examples are constructed using a general method in Chapter 2, each of them has a known eigenvalue and an eigenfunction, and these examples together form a rigorous general-purpose test of any such implementation.

Chapter 3 describes the numerical implementation of the index problem. It is well-known that the spectrum of the self-adjoint S-L problem solely contains real eigenvalues; if the leading coefficient function f is positive, i.e., $f > 0$ a.e. in (a,b), then these eigenvalues can be ordered to form a non-decreasing sequence

$$\lambda_1, \lambda_2, \lambda_3, \cdots \qquad (1.3.1)$$

approaching $+\infty$; if f changes sign, then the eigenvalues form a non-decreasing sequence

$$\cdots, \lambda_{-2}, \lambda_{-1}, \lambda_0, \lambda_1, \lambda_2, \cdots \qquad (1.3.2)$$

converging to both $-\infty$ and $+\infty$; in each sequence, the number of times an eigenvalue appears is equal to its (geometric) multiplicity. Here the indices of the eigenvalues in (1.3.2) are not determined by requiring λ_{-1} to be the negative eigenvalue closest to 0; instead, they are defined via the Prüfer angle method if the BC (boundary condition) is a separated one [97] and in terms of inequalities among eigenvalues if the BC is a coupled one [98]. (Even when f is positive, there are always negative eigenvalues for some self-adjoint BCs, but λ_{-1} never exists.) Given an eigenvalue λ_* of the problem, with a coupled BC, we deal with the following question: how to compute its index (if it is simple) or indices (if it is double) efficiently? This is the so-called index problem: in general, computing the index or indices by definitions or inequalities among eigenvalues is not efficient, since we do not have any estimate of the index or indices in advance. When f is positive, the number m of zeros of the real part or imaginary part of any eigenfunction for λ_* on $[a,b)$ provides such an estimate: the index or indices of λ_* is or are among $m-1$, m and $m+1$. See, for example, Theorem 4.8 of [48]. When f changes sign, we do not have such a general estimate. Note that even in the positive f case, the Prüfer angle method can not be directly used in general, since the eigenfunction is necessarily non-real when the BC is so. The index problem has practical importance: given a self-adjoint S-L problem, usually we can approximate the eigenvalues of the problem in a chosen interval, say $[-100,100]$, hence it is natural to ask what the indices of these eigenvalues are. Note that even when the spectrum is bounded from below, the indices are not obvious in general, since it is always possible (for example, for some coupled BCs) that the first few eigenvalues are in the left outside of the chosen interval, no matter how big the interval is. Fulton conjectured that for $k \in \mathbb{R}\backslash\{0\}$, $\lambda_* = 1$ is simple and its index is 2 [99], see Example 3.3.10. In [100], a simple solution to the index problem has been obtained, and it works in the same way for both the positive f case and the indefinite f case. As an application of the solution of the index problem, this conjecture has been proven to be true. More precisely, using the results on the level surfaces of the n-th eigenvalue and inequalities among eigenvalues for coupled BCs and those for separated BCs, separated BCs having λ_* also as an eigenvalue with the same index or indices are identified, and the index or indices are then determined by applying the Prüfer angle method once.

Chapter 4 relates the eigenvalues

$$\cdots < \lambda_{-2} < \lambda_{-1} < \lambda_0 < \lambda_1 < \lambda_2 < \cdots \tag{1.3.3}$$

of the problem

$$-(fy')' + qy = \lambda wy \text{ in } (a,b) \tag{1.3.4}$$

and

$$\cos \alpha \cdot y(a) - \sin \alpha \cdot (fy')(a) = 0 = \cos \beta \cdot y(b) - \sin \beta \cdot (fy')(b), \tag{1.3.5}$$

to the eigenvalues

$$\mu_0(\gamma) > \mu_{-1}(\gamma) > \mu_{-2}(\gamma) > \cdots \longrightarrow -\infty \tag{1.3.6}$$

and

$$\nu_0(\gamma) < \nu_1(\gamma) < \nu_2(\gamma) < \cdots \longrightarrow +\infty \tag{1.3.7}$$

of two corresponding one-parameter families of problems

$$-(fy')' + qy = \lambda wy \text{ in } (a,c), \tag{1.3.8}$$

$$\cos \alpha \cdot y(a) - \sin \alpha \cdot (fy')(a) = 0 = \cos \gamma \cdot y(c) - \sin \gamma \cdot (fy')(c) \tag{1.3.9}$$

and

$$-(fy')' + qy = \lambda wy \text{ in } (c,b), \tag{1.3.10}$$

$$\cos \gamma \cdot y(c) - \sin \gamma \cdot (fy')(c) = 0 = \cos \beta \cdot y(b) - \sin \beta \cdot (fy')(b) \tag{1.3.11}$$

with a definite f (i.e., f does not change the sign in each of the two subintervals above), where the parameter $\gamma \in [0,\pi)$. See [101].

First, we give a geometric characterization of the eigenvalues λ_n: they are the λ-coordinates of the intersections of the curves $\lambda = \mu_0(\gamma)$, $\lambda = \mu_{-1}(\gamma)$, $\lambda = \mu_{-2}(\gamma) \cdots$, and $\lambda = \nu_0(\gamma)$, $\lambda = \nu_1(\gamma)$, $\lambda = \nu_2(\gamma)$, $\cdots$ on $[0, \pi)$. This characterization yields simple proofs of the existence of the λ_n in [102] and their Prüfer angle characterization in [97]. It is also used to verify the way for determining the index n of each λ_n by a weighted count of the zeros of the eigenfunctions for λ_n in [98]. Then, we obtain interlacing relations among λ_n, $\mu_j(0)$ and $\nu_k(0)$. Such relations provide a means of studying the eigenvalues of relatively complicated S-L problems using the eigenvalues of simpler problems. For example, in terms of these relations, they obtain a simple proof of the asymptotic formulas for λ_n in [103], and explicitly describe the level surfaces of λ_n in [104]. The main idea of this chapter, i.e., the geometric characterization of the eigenvalues λ_n using simpler eigenvalues μ_j and ν_k, can be generalized to many other situations. So, in a certain sense, the major emphasis of this chapter is this new and geometric idea, rather than the results deduced from it.

Third-order differential equations arise in many physical phenomena, for example, three-layer beam problem, for more backgrounds we can refer to [71]. Hao characterized the self-adjoint domain of odd-order differential operators by using real parameter solutions [72], and Niu gave all canonical forms of self-adjoint BCs for regular third-order differential operators [73]. Uğurlu also studied a class of third-order self-adjoint boundary value problems and introduced the dependence of eigenvalues on the problem [76], and generalized these results to boundary value problems with discontinuity [77]. However, there is still lots of work that needs to be done for third-order differential operators, namely, differential operators with transmission conditions and eigenparameter-dependent BCs.

In Chapter 5, we investigate self-adjointness, the properties of eigenvalues, Green function of regular third-order differential operators with mixed and eigenparameter-dependent BCs, and consider third-order boundary value transmission conditions. By using the classical analysis techniques and spectral theory of linear operator, we transfer the BVTP to a self-adjoint operator T in an appropriate Hilbert space $\mathcal{H}$ such that the eigenvalues of the problem coincide with those of T.

In 1934, Paley and Wiener studied the problem of finding sequences $\{\lambda_n\}$ for which $\{\exp(i\lambda_n x)\}$ is a Riesz basis in $L^2[-\pi, \pi]$ [105]. Since then many results on the Riesz basis have been obtained [106–108]. Also, the Riesz basis of the systems of sines and cosines in $L^2[0, \pi]$, and the Riesz basis associated with S-L problems have been studied in many papers [109–118]. Moreover, on the problems of expansion of eigenfunctions, we refer to [119–

124].

In Chapter 6, We primarily focus on two applications of S-L problems. In section 6.1, we consider the problem of finding a new sequence associated with eigenfunctions of the S-L problem

$$\begin{cases} -y'' + qy = \lambda y, \text{ on } [0,\pi]; \\ y(0) = y(\pi) = 0, \end{cases} \quad (1.3.12)$$

such that it forms a Riesz basis. Section 6.2 discusses how to utilize S-L problems to describe the vertical structure of internal waves in the ocean.

Throughout this book unless noted otherwise, $L((a,b), \mathbb{R})$ denotes the space of real-valued Lebesgue integrable functions in (a,b), and SL(2,$\mathbb{R}$) will always denote the set of real matrices in dimension 2 and having determinant 1. We use $\boldsymbol{A}^C$ to denote the matrix of the cofactor of $\boldsymbol{A}$, and $\boldsymbol{A}^T$ stands for transpose of $\boldsymbol{A}$, while $\boldsymbol{A}^*$ is the complex conjugate transpose of $\boldsymbol{A}$. For $n, m \in \mathbb{N}$, we use $M_{n,m}(\mathbb{C})$ to denote the vector space of n by m complex matrices and $M_{n,m}(\mathbb{C}_*)$ the open subset of $M_{n,m}(\mathbb{C})$ consisting of the elements with the maximum rank $\min\{n,m\}$, the definitions of $M_{n,m}(\mathbb{R})$ and $M_{n,m}(\mathbb{R}_*)$ is similar. When a capital Latin or Greek letter stands for a matrix, the entries of the matrix are always denoted by the corresponding lowercase letter with two subindexes. For example, the entries of a matrix $\boldsymbol{K}$ are k_{ij}.

Chapter 2 Approximations of eigenvalues and eigenfunctions

In this chapter, we consider a regular S-L problem, i.e., the spectral problem consisting of the S-L equation

$$-(fy')' + qy = \lambda w y \text{ in } (a,b), \qquad (2.0.1)$$

and a regular boundary condition (BC), where

$$-\infty < a < b < +\infty,\ 1/f,\ q,\ w \in L((a,b), \mathbb{R}),\ w > 0 \text{ a.e., in } (a,b), \qquad (2.0.2)$$

and $\lambda \in \mathbb{C}$ is the so-called spectral parameter. Note that the leading coefficient function f is allowed to change the sign in (a,b), i.e., is indefinite: both $\{x \in (a,b);\ f(x) > 0\}$ and $\{x \in (a,b);\ f(x) < 0\}$ have a positive Lebesgue measure.

It is well-known that the spectrum of the problem can contain real or non-real eigenvalues.

2.1 Notation and theoretic results

In this section, we recall some basic results about two types of S-L problems and S-L equations with constant coefficient functions.

2.1.1 Usual S-L problems

By a solution to (2.0.1) we mean a function y in (a,b) such that y and fy' are absolutely continuous in all compact subintervals of (a,b) and (2.0.1) is satisfied almost everywhere (a.e.). The integrability conditions in (2.0.2) guarantee that for any solution y to (2.0.1), y and fy' have finite limits at both a and b. From now on, we will denote fy' by $y^{[1]}$ for any solution y to (2.0.1).

For every $\lambda \in \mathbb{C}$, let $\phi_{11}(\cdot, \lambda)$ and $\phi_{12}(\cdot, \lambda)$ be the solutions to (2.0.1) determined by the initial conditions

$$\phi_{11}(a, \lambda) = 1,\ \phi_{11}^{[1]}(a, \lambda) = 0,\ \phi_{12}(a, \lambda) = 0,\ \phi_{12}^{[1]}(a, \lambda) = 1. \qquad (2.1.1)$$

Then, they form a fundamental set of solutions to (2.0.1). We denote $\phi_{11}^{[1]}$ and $\phi_{12}^{[1]}$ by ϕ_{21} and ϕ_{22}, respectively. Set

$$\boldsymbol{\Phi}(x,\lambda) = \begin{pmatrix} \phi_{11}(x,\lambda) & \phi_{12}(x,\lambda) \\ \phi_{21}(x,\lambda) & \phi_{22}(x,\lambda) \end{pmatrix}, \quad x \in [a,b], \ \lambda \in \mathbb{C}. \tag{2.1.2}$$

Then, $\boldsymbol{\Phi}(x,\lambda)$ satisfies the matrix form of (2.0.1), i.e.,

$$\boldsymbol{\Phi}'(x,\lambda) = \begin{pmatrix} 0 & 1/f(x) \\ q(x) - \lambda w(x) & 0 \end{pmatrix} \boldsymbol{\Phi}(x,\lambda) \text{ in } (a,b), \tag{2.1.3}$$

and the initial condition $\boldsymbol{\Phi}(a,\lambda) = \boldsymbol{I}$. We call $\boldsymbol{\Phi}$ the fundamental solution matrix of (2.0.1). Note that $\boldsymbol{\Phi}(x,\lambda) \in \mathrm{SL}(2,\mathbb{R})$ for $x \in [a,b]$ and $\lambda \in \mathbb{R}$, and $\boldsymbol{\Phi}(x,\lambda) \in \mathrm{SL}(2,\mathbb{C})$ for $x \in [a,b]$ and $\lambda \in \mathbb{C}$.

For any solution y to (2.0.1), let

$$\boldsymbol{Y}(x) = \begin{pmatrix} y(x) \\ y^{[1]}(x) \end{pmatrix}, \quad x \in (a,b). \tag{2.1.4}$$

Then, BCs are specified by algebraic systems of the form

$$\boldsymbol{A}\boldsymbol{Y}(a) + \boldsymbol{B}\boldsymbol{Y}(b) = \boldsymbol{0}, \tag{2.1.5}$$

where $\boldsymbol{A}, \boldsymbol{B} \in M_{2,2}(\mathbb{C})$ such that $(\boldsymbol{A}\,|\,\boldsymbol{B}) \in M_{2,4}(\mathbb{C}_*)$. Note that equivalent algebraic equations of this form define the same BC. The BC (2.1.5) is said to be self-adjoint if

$$\boldsymbol{A}\begin{pmatrix} 0 & 1 \\ -1 & 0 \end{pmatrix}\boldsymbol{A}^* = \boldsymbol{B}\begin{pmatrix} 0 & 1 \\ -1 & 0 \end{pmatrix}\boldsymbol{B}^*. \tag{2.1.6}$$

Each separated self-adjoint BC has its standard form

$$\cos\alpha \cdot y(a) - \sin\alpha \cdot (fy')(a) = 0 = \cos\beta \cdot y(b) - \sin\beta \cdot (fy')(b) \tag{2.1.7}$$

with $\alpha \in [0,\pi)$ and $\beta \in (0,\pi]$.

Every coupled self-adjoint BC can be written in the form

$$\boldsymbol{Y}(b) = e^{i\gamma}\boldsymbol{K}\boldsymbol{Y}(a) \tag{2.1.8}$$

with $\gamma \in [0,\pi)$ and $\boldsymbol{K} \in \mathrm{SL}(2,\mathbb{R})$. Sometimes, it is convenient to allow other ranges of γ, such as $(-\pi,\pi)$ and $\mathbb{R}$.

A number $\lambda_* \in \mathbb{C}$ is called an eigenvalue of the S-L problems consisting of (2.0.1) and (2.1.5) if the S-L equation (2.0.1) with $\lambda = \lambda_*$ has a non-trivial solution satisfying the BC (2.1.5); and such a solution is called an eigenfunction for this eigenvalue. The vector space spanned by the eigenfunctions for an eigenvalue is the eigenspace for the eigenvalue, while the complex dimension of the eigenspace is called the geometric multiplicity of the eigenvalue.

The following result can be obtained by variation of parameters formula and Theorem of continuous dependence of solution on problem, see [3].

Theorem 2.1.1 Let $y = y(\cdot, \lambda)$ be a solution to (2.0.1) with $y(a, \lambda)$ and $(fy')(a, \lambda)$ independent λ, then on $[a, b] \times \mathbb{C}$,

$$\partial_\lambda y = \phi_{11}\gamma_2 - \phi_{12}\gamma_1, \quad \partial_\lambda(fy') = \phi_{21}\gamma_2 - \phi_{22}\gamma_1, \tag{2.1.9}$$

where

$$\gamma_i(x, \lambda) = \int_a^x \phi_{1i}(s, \lambda) y(s, \lambda) w(s) ds, \quad i = 1, 2. \tag{2.1.10}$$

Moreover, for each $x \in [a, b]$, $y(x, \lambda)$ and $(fy')(x, \lambda)$ are entire in λ.

The following formula can be verified directly using the ordinary differential equation about $\partial_\lambda \Phi(x, \lambda)$ derived from (2.1.3) and initial condition $\partial_\lambda \Phi(x, \lambda) = 0$, see [49].

Corollary 2.1.1 For each $x \in [a, b]$, $\Phi(x, \lambda)$ is an entire matrix function of λ, and

$$\partial_\lambda \Phi(x, \lambda) = \Phi(x, \lambda) \begin{pmatrix} a_{21}(x, \lambda) & a_{22}(x, \lambda) \\ -a_{11}(x, \lambda) & -a_{12}(x, \lambda) \end{pmatrix}, \tag{2.1.11}$$

where

$$a_{ij}(x, \lambda) = \int_a^x \phi_{1i}(s, \lambda) \phi_{1j}(s, \lambda) w(s) ds, \quad i, j = 1, 2. \tag{2.1.12}$$

The following result says that $\Phi(b, \lambda)$ characterizes the eigenvalues of the S-L problems.

Theorem 2.1.2 A number $\lambda \in \mathbb{C}$ is an eigenvalue of the S-L problem consisting of (2.0.1) and (2.1.5) if and only if

$$\Delta(\lambda) =: \det(A + B\Phi(b, \lambda)) = 0. \tag{2.1.13}$$

Proof. See [3]. □

By Corollary 2.1.1, the function $\Delta(\lambda)$ is entire in λ, to be called the characteristic function of the S-L problem, for its importance. Therefore, either all the complex numbers are eigenvalues, or the eigenvalues are isolated and do not have an accumulation point in $\mathbb{C}$. We remark that after row operations are applied to $(A \mid B)$, $\Delta(\lambda)$ may gain a non-zero constant factor.

Note that a number $\lambda_* \in \mathbb{C}$ is an eigenvalue for (2.1.5) of geometric multiplicity 1 if and only if $A + B\Phi(b, \lambda_*)$ has rank 1; in this case, if $(d_1 \; d_2)$ is a non-zero row of $A + B\Phi(b, \lambda_*)$, then $d_2\phi_{11}(\cdot, \lambda_*) - d_1\phi_{12}(\cdot, \lambda_*)$ is an eigenfunction for λ_*. When $\lambda_* \in \mathbb{C}$ is an eigenvalue of geometric multiplicity 2, any non-trivial linear combination $c_1\phi_{11}(\cdot, \lambda_*) + c_2\phi_{12}(\cdot, \lambda_*)$ is an eigenfunction for λ_*.

The formula in the next result comes from straightforward computations.

Proposition 2.1.1 The characteristic function for the S-L problem consisting of (2.0.1) and (2.1.5) is given by

$$\Delta(\lambda) = \det A + \det B + \sum_{i,j=1}^{2} c_{ij}\phi_{ij}(b, \lambda), \tag{2.1.14}$$

where

$$\begin{pmatrix} c_{12} & c_{22} \\ c_{21} & c_{22} \end{pmatrix} = B^T A^C = \begin{pmatrix} b_{11} & b_{21} \\ b_{12} & b_{22} \end{pmatrix} \begin{pmatrix} a_{22} & -a_{21} \\ -a_{12} & a_{11} \end{pmatrix}. \tag{2.1.15}$$

The analytic multiplicity (or just multiplicity) of an isolated eigenvalue is its order as a zero of Δ. An eigenvalue is said to be simple if it has multiplicity 1, while the eigenvalues of multiplicity 2 are called double eigenvalues. When we count the (isolated) eigenvalues of an S-L problem in a domain in $\mathbb{C}$, their multiplicities will be taken into account.

The reality of f, q in (2.0.1) and $\Phi(b, \lambda)$ for $\lambda \in \mathbb{R}$ implies the following result, see [49].

Lemma 2.1.1 The non-real eigenvalues for a real boundary condition appear in conjugate pairs. Each such pair shares the same multiplicity and the same geometric multiplicity.

The following result is well known, see [94].

Theorem 2.1.3 If f and $w > 0$ a.e. in (a, b) and the boundary condition (2.1.5) is self-adjoint, then the eigenvalues of the S-L problem consisting of (2.0.1) and (2.1.5) are all real.

The following result has appeared in [69].

Theorem 2.1.4 If f changes the sign in (a, b) and $w > 0$ a.e. in (a, b), then the eigenvalues of the S-L problem consisting of (2.0.1) and self-adjoint boundary condition are all real.

2.1.2 S-L problems with spectral-parameter-dependent BCs

A spectral-parameter-dependent boundary condition (λ-dependent BC) is a relation specified by an algebraic equation of the form

$$A(\lambda)Y(a) + B(\lambda)Y(b) = 0, \tag{2.1.16}$$

where $A(\lambda)$ and $B(\lambda)$ are 2×2 matrices analytic on a domain $\tilde{R}$ in $\mathbb{C}$, and the matrix $(A(\lambda) | B(\lambda))$ has two columns such that their determinant is not identically zero. Note that equivalent algebraic equations of this form define the same BC. The matrix $(A(\lambda) | B(\lambda))$ is called "the" coefficient matrix of the BC specified by (2.1.16). Here the quotation mark is used to indicate that $(A(\lambda) | B(\lambda))$ is unique only up to a left factor which is analytic on $\tilde{R}$ and whose determinant is never zero in $\tilde{R}$.

A number $\lambda_* \in \tilde{R}$ is called an eigenvalue of the S-L problem consisting of (2.0.1) and (2.1.16) if the S-L equation (2.0.1) with $\lambda = \lambda_*$ has a nontrivial solution satisfying the BC (2.1.16) with $\lambda = \lambda_*$; and such a solution is called an eigenfunction for this eigenvalue. The definitions of the eigenspace and the geometric multiplicity are similar to those for usual S-L problems.

An almost trivial way for obtaining λ-dependent BCs is multiplying a column or a row of the coefficient matrices of usual BCs by a function $h(\lambda)$ that is analytic on $\tilde{R}$ and is not identically constant. The eigenvalues for such a λ-dependent BC consist of the zeros of $h(\lambda)$ in $\tilde{R}$ and the eigenvalues for the corresponding usual BC in $\tilde{R}$.

When $\tilde{R}$ is not given, we always assume that $\tilde{R}$ equals $\mathbb{C}$, unless another choice for $\tilde{R}$ is obvious and better.

Note that the fundamental matrix $\Phi(x, \lambda)$ of an S-L equation has been given above. The following two results are similar to the corresponding ones for usual S-L problems.

Theorem 2.1.5 A number $\lambda \in \mathbb{C}$ is an eigenvalue of the S-L problem consisting of (2.0.1) and (2.1.16) if and only if

$$\Delta(\lambda) =: \det(A(\lambda) + B(\lambda)\Phi(b, \lambda)) = 0. \tag{2.1.17}$$

The function $\Delta(\lambda)$ is analytic in λ on $\tilde{R}$, to be also called the characteristic function of the S-L problem.

Proposition 2.1.2 The characteristic function of the S-L problem consisting of (2.0.1) and (2.1.16) is given by

$$\Delta(\lambda) = \det A(\lambda) + \det B(\lambda) + \sum_{i,j=1}^{2} c_{ij}(\lambda)\phi_{ij}(b, \lambda), \tag{2.1.18}$$

where

$$C(\lambda) = B(\lambda)^T A(\lambda)^C = \begin{pmatrix} b_{11}(\lambda) & b_{21}(\lambda) \\ b_{12}(\lambda) & b_{22}(\lambda) \end{pmatrix} \begin{pmatrix} a_{22}(\lambda) & -a_{21}(\lambda) \\ -a_{12}(\lambda) & a_{11}(\lambda) \end{pmatrix}. \tag{2.1.19}$$

2.1.3 S-L equations with piece-wise constant coefficients

If there are $n \in \mathbb{N}$ and $a_0 = a, a_1, \cdots, a_{n-1}, a_n = b \in (-\infty, \infty)$ in strictly increasing order such that f, q, and w are constant in (a_{i-1}, a_i) for $i = 1, 2, \cdots, n$, then we say that the S-L equation (2.0.1) has piece-wise constant coefficient functions. In this case, since the constant values of $1/f$ in (a, a_1) and (a_{n-1}, b) are non-zero, we must have $a, b \in \mathbb{R}$. We want to find the fundamental matrix of such S-L equations. As the outcome of the easiest case, we have the following result, which generalizes the fundamental matrix of the Fourier equation.

The following two results see [3].

Proposition 2.1.3 If f, q, and w are constant in (a, b), then a and b are finite, and the fundamental matrix of (2.0.1) is given by

$$\Phi(x, \lambda) = \begin{pmatrix} \cosh p(x, \lambda) & \dfrac{1}{g(\lambda)} \sinh p(x, \lambda) \\ g(\lambda) \sinh p(x, \lambda) & \cosh p(x, \lambda) \end{pmatrix}, \quad x \in [a, b], \lambda \in \mathbb{C}, \quad (2.1.20)$$

where

$$p(x, \lambda) = (x - a)\sqrt{(q - w\lambda)/f}, \quad g(\lambda) = \sqrt{f(q - w\lambda)}. \quad (2.1.21)$$

Note that

$$\frac{1}{g(\lambda)} \sinh p(x, \lambda) = \frac{x - a}{f} + \frac{(x - a)^3}{3! f^2}(q - w\lambda) + \frac{(x - a)^5}{5! f^3}(q - w\lambda)^2 + \cdots \quad (2.1.22)$$

is entire on $\mathbb{C}$. The next result and Proposition 2.1.3 imply a way for computing the fundamental matrix of an S-L equation with piece-wise constant coefficient functions.

Proposition 2.1.4 Let $-\infty < a_0 < a_1 < a_2 < \infty$. Assume that Φ_i is the fundamental matrix of

$$-(f_i y')' + q_i y = \lambda w_i y \quad \text{in } (a_{i-1}, a_i) \quad (2.1.23)$$

for $i = 1$ and 2, then

$$\Phi(x, \lambda) = \begin{cases} \Phi_1(x, \lambda) & \text{for } x \in [a_0, a_1], \lambda \in \mathbb{C}, \\ \Phi_2(x, \lambda)\Phi_1(a_1, \lambda) & \text{for } x \in (a_1, a_2], \lambda \in \mathbb{C} \end{cases} \quad (2.1.24)$$

is the fundamental matrix of

$$-(fy')' + qy = \lambda wy \quad \text{in } (a_0, a_2), \quad (2.1.25)$$

where

$$f(x) = \begin{cases} f_1(x) & \text{if } x \in [a_0, a_1], \\ f_2(x) & \text{if } x \in (a_1, a_2], \end{cases} \qquad (2.1.26)$$

while q_i and w_i are defined similarly.

2.2 Main ideas of the algorithms

In this section, we recall a special case of the argument principle in complex analysis, and describe the main ideas of the algorithm.

Theorem 2.2.1 Assume that $R \subset \mathbb{C}$ is a bounded region whose boundary ∂R consists of rectifiable curves. If a function $\psi(z)$ is analytic in z on the domain R, and ∂R does not contain any zeros of $\psi(z)$, then

$$\frac{1}{2\pi i} \int_{\partial R} \frac{\psi'(z)}{\psi(z)} dz = \text{the number of zeros of } \psi(z) \text{ inside } R. \qquad (2.2.1)$$

Remark 2.2.1 In fact,

$$\frac{1}{i} \int_{\partial R} \frac{\psi'(z)}{\psi(z)} dz = \text{change in arg } \psi(z) \text{ along } \partial R. \qquad (2.2.2)$$

Therefore, when $\psi(z)$ is explicit, we can compute the number of zeros of $\psi(z)$ inside R using (2.2.1); when $\psi(z)$ is not explicitly available, we usually compute the change in arg $\psi(z)$ along ∂R directly, i.e., using the values of $\psi(z)$ on ∂R.

Given an S-L problem and a rectangle region R, neither Theorem 2.2.1 or Remark 2.2.1 yields the following algorithm for approximating the eigenvalues of the problem in the region, together with their multiplicities, and the corresponding eigenfunctions.

Step 1. If the S-L equation of the S-L problem has piecewise constant functions, then compute $\Phi(x, \lambda)$ explicitly using Proposition 2.1.3 and Proposition 2.1.4. For a usual S-L problem, take suitable care of the endpoints a and b when some coefficient functions of (2.0.1) are unbounded at the endpoints a and b, the corresponding approximations of a and b are denoted by aa and bb, respectively. Set up the ODE (ordinary differential equation) solver so that later whenever a value of λ is given, an approximation of the fundamental matrix $\Phi(x, \lambda)$ for $aa \leqslant x \leqslant bb$ can be computed using

$$\Phi'(x, \lambda) = \begin{pmatrix} 0 & 1/f(x) \\ q(x) - \lambda w(x) & 0 \end{pmatrix} \Phi(x, \lambda) \quad \text{in } (aa, bb), \quad \Phi(aa, \lambda) \equiv \boldsymbol{I}.$$

$$(2.2.3)$$

Step 2. Obtain an expression for the characteristic function $\Delta(\lambda)$ using Proposition 2.1.1 in the case of a λ-independent BC, Proposition 2.1.2 in the case of a λ-dependent BC.

Step 3. Count the total number ne of zeros of $\Delta(\lambda)$ (i.e., the eigenvalues of the S-L problem in our case) in R in terms of either Theorem 2.2.1 or Remark 2.2.1. If $ne=0$, the execution of the codes ends. If $ne=1$, find an approximation of the eigenvalue in the region using the Newton iteration method; in the situation where the Newton iteration method fails to find an approximation, then adopt the method for $ne >1$ to be followed. If $ne >1$, we have two subcases to handle: when the region is small enough, choose an appropriate point in the region to serve as an approximation of the eigenvalues in the region, and ne is regarded as the multiplicity of the approximate eigenvalue; otherwise, divide the region into smaller regions, keep track of the number of zeros of $\Delta(\lambda)$ inside each smaller region by either Theorem 2.2.1 or Remark 2.2.1, throw away the smaller regions not containing any zeros of $\Delta(\lambda)$, and treat each remaining smaller region as above. This region-dividing procedure is applied enough times, until approximations of all eigenvalues of the problem in R, with the given accuracy are found.

Step 4. When λ_* is an eigenvalue of geometric multiplicity 1 and (d_1, d_2) is a non-zero row of $\boldsymbol{A} + \boldsymbol{B}\boldsymbol{\Phi}(b, \lambda_*)$ in the case of a λ-independent BC, $\boldsymbol{A}(\lambda) + \boldsymbol{B}(\lambda)\boldsymbol{\Phi}(b, \lambda_*)$ in the case of a λ-dependent BC, $d_2\phi_{11}(\cdot, \lambda) - d_1\phi_{12}(\cdot, \lambda)$ is an eigenfunction for λ_*; when λ_* is an eigenvalue of geometric multiplicity 2, $\phi_{11}(\cdot, \lambda)$ and $\phi_{12}(\cdot, \lambda)$ are eigenfunctions for λ_*.

2.3 General methods for constructing examples

In this section, we give general ways of constructing S-L problems with known eigenvalues and eigenfunctions. These S-L problems include usual S-L problems and S-L problems with λ-dependent BCs.

2.3.1 Usual S-L problems

The following is a general way for constructing usual S-L problems with known eigenvalues and eigenfunctions.

Theorem 2.3.1 Let $1/f, w \in L((a,b), \mathbb{C})$, and $\lambda_* \in \mathbb{C}$. If $y_* \in \mathrm{AC}_{\mathrm{loc}}((a,b), \mathbb{C}\setminus\{0\})$ satisfies the conditions that $fy'_* \in \mathrm{AC}_{\mathrm{loc}}((a,b), \mathbb{C})$ and $(fy'_*)'/y_* \in L((a,b), \mathbb{C})$, then

$$q := \lambda_* w + (fy'_*)'/y_* \qquad (2.3.1)$$

is in $L((a,b),\mathbb{C})$, and λ_* is an eigenvalue of the S-L problem

$$\begin{cases} -(fy')' + qy = \lambda wy \text{ in } (a,b), \\ c_2 y(a) - c_1 (fy')(a) = 0 = d_2 y(b) - d_1 (fy')(b) \end{cases} \quad (2.3.2)$$

with an eigenfunction y_*, where

$$c_1 = y_*(a), \quad c_2 = (fy'_*)(a), \quad d_1 = y_*(b), \quad d_2 = (fy'_*)(b). \quad (2.3.3)$$

Note that the DE (differential equation) in the S-L problem (2.3.2) is regular, and hence its solution y_* and the quasi-derivative fy'_* of y_* have finite limits at both a and b, i.e., the constants c_1, c_2, d_1 and d_2 are well-defined. Moreover, one of c_1 and c_2 is non-zero since y_* is a non-trivial solution to the DE in (2.3.2), and similarly for d_1 and d_2.

The BC in (2.3.2) is the only separated BC satisfied by y_*; however, there are many coupled BCs fulfilled by y_*. The BCs satisfied by y_* can be grouped into six overlapping families, each of which has two free complex parameters when it is not empty. The following proposition gives all these BCs.

Proposition 2.3.1 The above function y_* satisfies the following boundary conditions

$$\begin{pmatrix} 1 & 0 & b_{11} & b_{12} \\ 0 & 1 & b_{21} & b_{22} \end{pmatrix} \begin{pmatrix} Y(a) \\ Y(b) \end{pmatrix} = \mathbf{0}, \quad (2.3.4)$$

when one of b_{11}, b_{12}, b_{21} and b_{22} is non-zero, and either

$$b_{j1} = -(c_j + b_{j2}d_2)/d_1 \quad \text{for } j = 1,2 \quad \text{if } d_1 \neq 0, \quad \text{or} \quad (2.3.5)$$

$$b_{j2} = -(c_j + b_{j1}d_1)/d_2 \quad \text{for } j = 1,2 \quad \text{if } d_2 \neq 0; \quad (2.3.6)$$

$$\begin{pmatrix} 1 & a_{12} & 0 & b_{12} \\ 0 & a_{22} & -1 & b_{22} \end{pmatrix} \begin{pmatrix} Y(a) \\ Y(b) \end{pmatrix} = \mathbf{0}, \quad (2.3.7)$$

when one of c_2, d_2 is non-zero, and either

$$a_{12} = -(c_1 + b_{12}d_2)/c_2, \quad a_{22} = (d_1 - b_{22}d_2)/c_2 \quad \text{if } c_2 \neq 0, \quad \text{or} \quad (2.3.8)$$

$$b_{12} = -(c_1 + a_{12}c_2)/d_2, \quad b_{22} = (d_1 - a_{22}c_2)/d_2 \quad \text{if } d_2 \neq 0; \quad (2.3.9)$$

$$\begin{pmatrix} 1 & a_{12} & b_{11} & 0 \\ 0 & a_{22} & b_{21} & -1 \end{pmatrix} \begin{pmatrix} Y(a) \\ Y(b) \end{pmatrix} = \mathbf{0}, \quad (2.3.10)$$

when one of c_2, d_1 is non-zero, and either

$$a_{12} = -(c_1 + b_{11}d_1)/c_2, \quad a_{22} = (d_2 - b_{21}d_1)/c_2 \quad \text{if } c_2 \neq 0, \quad \text{or} \quad (2.3.11)$$

$$b_{11} = -(c_1 + a_{12}c_2)/d_1, \quad b_{21} = (d_2 - a_{22}c_2)/d_1 \quad \text{if } d_1 \neq 0; \tag{2.3.12}$$

$$\begin{pmatrix} a_{11} & 1 & 0 & b_{12} \\ a_{21} & 0 & -1 & b_{22} \end{pmatrix} \begin{pmatrix} Y(a) \\ Y(b) \end{pmatrix} = \mathbf{0}, \tag{2.3.13}$$

when one of c_1, d_2 is non-zero, and either

$$a_{11} = -(c_2 + b_{12}d_2)/c_1, \quad a_{21} = (d_1 - b_{22}d_2)/c_1 \quad \text{if } c_1 \neq 0, \quad \text{or} \tag{2.3.14}$$

$$b_{12} = -(a_{11}c_1 + c_2)/d_2, \quad b_{22} = (d_1 - a_{21}c_1)/d_2 \quad \text{if } d_2 \neq 0; \tag{2.3.15}$$

$$\begin{pmatrix} a_{11} & 1 & b_{11} & 0 \\ a_{21} & 0 & b_{21} & -1 \end{pmatrix} \begin{pmatrix} Y(a) \\ Y(b) \end{pmatrix} = \mathbf{0}, \tag{2.3.16}$$

when one of c_1, d_1 is non-zero, and either

$$a_{11} = -(c_2 + b_{11}d_1)/c_1, \quad a_{21} = (d_2 - b_{21}d_1)/c_1 \quad \text{if } c_1 \neq 0, \quad \text{or} \tag{2.3.17}$$

$$b_{11} = -(a_{11}c_1 + c_2)/d_1, \quad b_{21} = (d_2 - a_{21}c_1)/d_1 \quad \text{if } d_1 \neq 0; \tag{2.3.18}$$

$$\begin{pmatrix} a_{11} & a_{12} & -1 & 0 \\ a_{21} & a_{22} & 0 & -1 \end{pmatrix} \begin{pmatrix} Y(a) \\ Y(b) \end{pmatrix} = \mathbf{0}, \tag{2.3.19}$$

when one of a_{11}, a_{12}, a_{21} and a_{22} is non-zero, and either

$$a_{j1} = (d_j - a_{j2}c_2)/c_1 \quad \text{for } j = 1, 2 \quad \text{if } c_1 \neq 0, \quad \text{or} \tag{2.3.20}$$

$$a_{j2} = (d_j - a_{j1}c_j)/c_2 \quad \text{for } j = 1, 2 \quad \text{if } c_2 \neq 0. \tag{2.3.21}$$

Note that when $c_2 = d_2 = 0$ (resp., $c_2 = d_1 = 0$, $c_1 = d_2 = 0$, $c_1 = d_1 = 0$), the second (resp., third, fourth, fifth) family is empty. If $d_2 = 0$, then b_{11} and b_{21} given by (2.3.5) in the first family are constant and independent of the free parameters b_{12} and b_{22}; similarly for the other situations.

Remark 2.3.1 The self-adjoint BCs satisfied by y_* can be determined using (2.1.6) and Proposition 2.3.1.

2.3.2 S-L problems with λ-dependent BCs

The following is another key theorem of this section.

Theorem 2.3.2 Let $\lambda_* \in \mathbb{C}$, and $\boldsymbol{H}(\lambda)$ and $\boldsymbol{L}(\lambda)$ be 2×2 matrices analytic in λ on a neighborhood of λ_* in $\mathbb{C}$ such that

$$\boldsymbol{H}(\lambda_*) = \boldsymbol{L}(\lambda_*) = \mathbf{0}. \tag{2.3.22}$$

Then, λ_* is an eigenvalue of the S-L problem

$$\begin{cases} -(fy')' + qy = \lambda wy \text{ in } (a,b), \\ A(\lambda)Y(a) + B(\lambda)Y(b) = 0 \end{cases} \quad (2.3.23)$$

if and only if λ_* is an eigenvalue of the S-L problem

$$\begin{cases} -(fy')' + qy = \lambda wy \text{ in } (a,b), \\ (A(\lambda) + H(\lambda))Y(a) + (B(\lambda) + L(\lambda))Y(b) = 0. \end{cases} \quad (2.3.24)$$

In this case, the eigenspace for λ_* as an eigenvalue of (2.3.17) is equal to that of (2.3.18).

Proof. Note that the two problems share the same DE. So we only need to pay attention to BCs. For a solution y_* to the DE, set $\boldsymbol{Y}_* = (y_*, fy'_*)^T$.

If λ_* is an eigenvalue of (2.3.17), with an eigenfunction y_*, then

$$\boldsymbol{A}(\lambda_*)\boldsymbol{Y}_*(a) + \boldsymbol{B}(\lambda_*)\boldsymbol{Y}_*(b) = \boldsymbol{0}. \quad (2.3.25)$$

Thus, by (2.3.16),

$$(\boldsymbol{A}(\lambda_*) + \boldsymbol{H}(\lambda_*))\boldsymbol{Y}_*(a) + (\boldsymbol{B}(\lambda_*) + \boldsymbol{L}(\lambda_*))\boldsymbol{Y}_*(b) \\ = \boldsymbol{A}(\lambda_*)\boldsymbol{Y}_*(a) + \boldsymbol{B}(\lambda_*)\boldsymbol{Y}_*(b) = \boldsymbol{0}. \quad (2.3.26)$$

Hence, λ_* is also an eigenvalue of (2.3.18), with an eigenfunction y_*.

Conversely, if λ_* is an eigenvalue of (2.3.18), with an eigenfunction y_*, then

$$(\boldsymbol{A}(\lambda_*) + \boldsymbol{H}(\lambda_*))\boldsymbol{Y}_*(a) + (\boldsymbol{B}(\lambda_*) + \boldsymbol{L}(\lambda_*))\boldsymbol{Y}_*(b)) = \boldsymbol{0}. \quad (2.3.27)$$

Thus, by (2.3.16) again,

$$\boldsymbol{A}(\lambda_*)\boldsymbol{Y}_*(a) + \boldsymbol{B}(\lambda_*)\boldsymbol{Y}_*(b) \\ = (\boldsymbol{A}(\lambda_*) + \boldsymbol{H}(\lambda_*))\boldsymbol{Y}_*(a) + (\boldsymbol{B}(\lambda_*) + \boldsymbol{L}(\lambda_*))\boldsymbol{Y}_*(b) = \boldsymbol{0}. \quad (2.3.28)$$

Therefore, λ_* is also an eigenvalue of (2.3.17), with an eigenfunction y_*. □

A special case of Theorem 2.3.2 is the following result, which implies a general way of constructing S-L problems with λ-dependent BCs from usual S-L problems.

Corollary 2.3.1 Let λ_*, $\boldsymbol{H}(\lambda)$ and $\boldsymbol{L}(\lambda)$ be as in Theorem 2.3.2. Then, λ_* is an eigenvalue of the S-L problem

$$\begin{cases} -(fy')' + qy = \lambda wy \text{ in } (a,b), \\ AY(a) + BY(b) = 0 \end{cases} \quad (2.3.29)$$

if and only if λ_* is an eigenvalue of the S-L problem

$$\begin{cases} -(fy')' + qy = \lambda wy \text{ in } (a,b), \\ (\boldsymbol{A} + \boldsymbol{H}(\lambda))\boldsymbol{Y}(a) + (\boldsymbol{B} + \boldsymbol{L}(\lambda))\boldsymbol{Y}(b) = \boldsymbol{0}. \end{cases} \quad (2.3.30)$$

In this case, the eigenspace for λ_* as an eigenvalue of (2.3.29) is equal to that of (2.3.30).

Remark 2.3.2 Even though the BC in the S-L problem (2.3.29) does not involve λ, the BC in the S-L problem (2.3.30) is λ-dependent unless $\boldsymbol{H}(\lambda) \equiv \boldsymbol{L}(\lambda) \equiv 0$.

Another corollary of Theorem 2.3.2 is the following characterization of the eigenvalues of S-L problems with λ-dependent BCs using eigenvalues of usual S-L problems.

Corollary 2.3.2 A number $\lambda_* \in \mathbb{C}$ is an eigenvalue of the S-L problem

$$\begin{cases} -(fy')' + qy = \lambda wy \text{ in } (a,b), \\ \boldsymbol{A}(\lambda)\boldsymbol{Y}(a) + \boldsymbol{B}(\lambda)\boldsymbol{Y}(b) = \boldsymbol{0} \end{cases} \quad (2.3.31)$$

if and only if either

$$\text{rank}(\boldsymbol{A}(\lambda_*) \mid \boldsymbol{B}(\lambda_*)) \leqslant 1, \quad (2.3.32)$$

or $\text{rank}(\boldsymbol{A}(\lambda_*) \mid \boldsymbol{B}(\lambda_*)) = 2$ and λ_* is an eigenvalue of the S-L problem

$$\begin{cases} -(fy')' + qy = \lambda wy \text{ in } (a,b), \\ \boldsymbol{A}(\lambda_*)\boldsymbol{Y}(a) + \boldsymbol{B}(\lambda_*)\boldsymbol{Y}(b) = \boldsymbol{0}. \end{cases} \quad (2.3.33)$$

We would like to point out that (2.3.33) is a usual S-L problem, i.e., its BC does not involve λ.

2.4 Examples with λ-independent BCs

In this section, using Theorem 2.3.2 and Corollary 2.3.1, we construct some examples, then present approximations of the known eigenvalues and eigenfunctions given by codes. In all these examples, we will always use S-L problems with an exact eigenvalue 1 or i, this choice of the exact eigenvalue is not special: it can be shifted to any real number by adding a corresponding multiple of w to q.

Example 2.4.1 Let $q(x) = 1 + 2/(x^2 + 1)$. Then, 1 is an eigenvalue of the self-adjoint S-L problem

$$-y'' + qy = \lambda y \text{ in } (0, 1), \qquad y'(0) = 0 = y(1) - y'(1) \qquad (2.4.1)$$

with an eigenfunction $y(x) = x^2 + 1$. In this example, the BC is real, separated, and hence self-adjoint.

The codes with different values of accuracy yield the following approximations of the eigenvalue 1 on the interval $[-2, 2]$:

accuracy	approximation	
0.001	1.0004572	
0.0001	1.0000508	(2.4.2)
0.00001	1.0000056	
0.000001	1.0000006	

The above results indicate that the codes are able to obtain desired approximations for different given accuracies, assuming that the ODE solver in mathematica is sufficiently effective. The following graph (Figure 2.4.1) illustrates the eigenfunction $y(x) = x^2 + 1$ for $\lambda=1$ and its approximation so produced by the codes. In this book, numerical solutions are always graphed in the dotted line style.

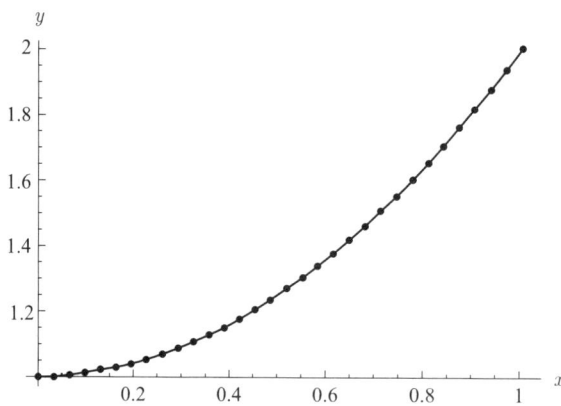

Figure 2.4.1 The eigenfunction and its approximation

In the remaining examples, unless otherwise stated, the parameter gaccuracy in codes takes the default value 0.0001.

Example 2.4.2 The number 1 is an eigenvalue of the S-L problem

$$-y'' + y = \lambda y \text{ in } (0, 1), \qquad iy(0) + y'(0) = 0 = y(1) - (1+i)y'(1) \qquad (2.4.3)$$

with an eigenfunction $y(x) = x + \mathrm{i}$. In this example, the BC is non-real, separated, and hence non-self-adjoint.

A search on the square $[-2, 100] \times [-2, 100]$ gives the simple eigenvalues

$$1., \ 9.8791 - 1.1608\mathrm{i}, \ 39.4752 - 1.0355\mathrm{i}, \ 88.8247 - 1.0153\mathrm{i}. \tag{2.4.4}$$

The following graphs (Figure 2.4.2) illustrate the real and imaginary parts of the eigenfunction $y(x) = x + \mathrm{i}$ for $\lambda = 1$ and their approximations.

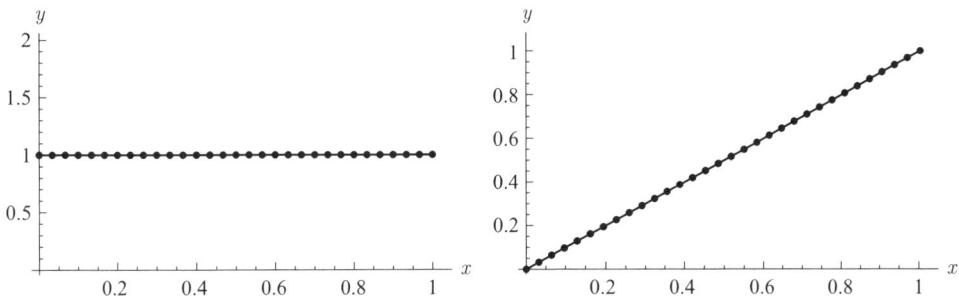

Figure 2.4.2 The real and imaginary parts of the eigenfunction and their approximations

Example 2.4.3 Let $q(x) = 1 + 2/(x^2 + 1)$. Then, 1 is an eigenvalue of the self-adjoint S-L problem

$$-y'' + qy = \lambda y \ \text{in} \ (0, 1), \quad \begin{pmatrix} 1 & 0 & -1/2 & 0 \\ 0 & -1/2 & -1 & 1 \end{pmatrix} \begin{pmatrix} \mathbf{Y}(0) \\ \mathbf{Y}(1) \end{pmatrix} = \mathbf{0} \tag{2.4.5}$$

with an eigenfunction $y(x) = x^2 + 1$. In this example, the BC is real, coupled, and self-adjoint.

A search on the interval $[-2, 200]$ yields the simple eigenvalue

$$1., \ 32.5399, \ 49.1352, \ 143.0192, \ 175.5661. \tag{2.4.6}$$

Example 2.4.4 The number 1 is an eigenvalue of the self-adjoint S-L problem

$$-y'' + y = \lambda y \ \text{in} \ (0, 1), \quad \begin{pmatrix} 1 & 0 & 0 & -\mathrm{i} \\ 0 & \mathrm{i} & -1 & 1 \end{pmatrix} \begin{pmatrix} \mathbf{Y}(0) \\ \mathbf{Y}(1) \end{pmatrix} = \mathbf{0} \tag{2.4.7}$$

with an eigenfunction $y(x) = x + \mathrm{i}$. In this example, the BC is non-real, coupled, and self-adjoint.

A search on the interval $[-2, 200]$ yields the simple eigenvalue

$$1., \ 12.5982, \ 42.3937, \ 91.7869, \ 160.8913. \tag{2.4.8}$$

Example 2.4.5 Let $q(x) = 1 + 2/(x^2 + 1)$. Then, $\lambda = 1$ is an eigenvalue of the

S-L problem

$$-y'' + qy = \lambda y \text{ in } (0, 1), \quad \begin{pmatrix} 1 & 0 & -1/2 & 0 \\ 0 & 1 & -1 & 1 \end{pmatrix} \begin{pmatrix} \mathbf{Y}(0) \\ \mathbf{Y}(1) \end{pmatrix} = \mathbf{0} \quad (2.4.9)$$

with an eigenfunction $y(x) = x^2 + 1$. In this example, the BC is real, coupled, and non-self-adjoint.

A search on the square $[-2, 100] \times [-2, 100]$ gives the simple eigenvalue

$$1., \ 13.2159, \ 82.6568, \ 92.0961. \quad (2.4.10)$$

Example 2.4.6 The number 1 is an eigenvalue of the S-L problem

$$-y'' + y = \lambda y \text{ in } (0, 1), \quad \begin{pmatrix} 1 & 0 & 0 & -i \\ 0 & i+1 & -1 & 0 \end{pmatrix} \begin{pmatrix} \mathbf{Y}(0) \\ \mathbf{Y}(1) \end{pmatrix} = \mathbf{0} \quad (2.4.11)$$

with an eigenfunction $y(x) = x + i$. In this example, the BC is non-real, coupled, and non-self-adjoint.

A search on the square $[-2, 100] \times [-2, 100]$ yields the simple eigenvalue

$$1., \ 9.8724 - 0.9526i, \ 41.4751 + 0.9657i, \ 88.8273 - 0.9952i. \quad (2.4.12)$$

Example 2.4.7 Let $k \in \mathbb{C}$, and define $q(x) = k^2 \cos^2 x - 3k \sin x$. Then, $\lambda = 1$ is an eigenvalue of the S-L problem

$$-y'' + qy = \lambda y \text{ in } (0, 2\pi), \quad \mathbf{Y}(2\pi) = \mathbf{Y}(0) \quad (2.4.13)$$

with an eigenfunction $y(x) = e^{k \sin x} \cos x$. This problem is self-adjoint when $k \in \mathbb{R}$.

For several values of k, a search on the interval $[-2, 2]$ gets the following eigenvalues, with their multiplicities in front of them

k	multiplicities and approximations
-2	1, 1.00005.
-1	1, -1.56155; 1, 1..
0	1, 0.; 2, 1.00005.
1	1, -1.56155; 1, 1..
2	1, 1.00005.

$$(2.4.14)$$

Figure 2.4.3 illustrates the eigenfunction $y(x) = e^{k \sin x} \cos x$ for $\lambda = 1$ and its approximation when $k = 2$.

By Theorem 2.2 in [47], for each $n \in \mathbb{N}$, the n-th eigenvalue of a self-adjoint S-L problem depends continuously on the S-L equation in the problem. The Figure 2.4.4 shows how eigenvalues of (2.4.13) vary when the parameter k in the S-L equation in (2.4.13) changes continuously.

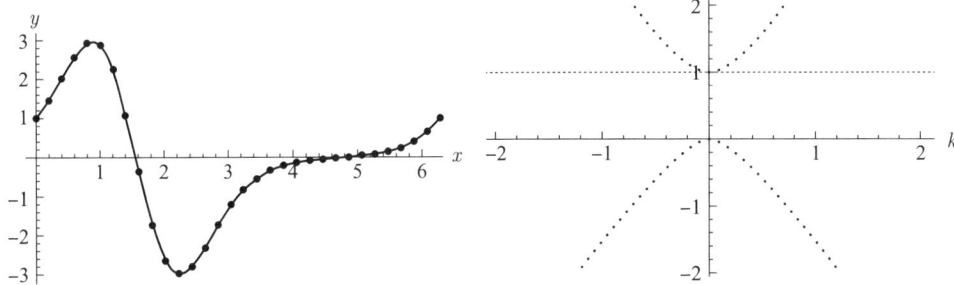

Figure 2.4.3 The eigenfunction and its approximation when $k = 2$

Figure 2.4.4 The varieties of the eigenvalues when k changes continuously

Example 2.4.8 For $\sigma, \tau \in \mathbb{R}$, let

$$f(x) = (1-x)^{\sigma+1}(1+x)^{\tau+1}, \qquad w(x) = (1-x)^{\sigma}(1+x)^{\tau}. \tag{2.4.15}$$

It is known that when $\sigma, \tau \in (-1, 0)$, the eigenvalues of the Neumann problem

$$-(fy')' = \lambda wy \text{ in } (-1, 1), \qquad (fy')(-1) = 0 = (fy')(1) \tag{2.4.16}$$

associated with the Jacobi equation are

$$\lambda_n = (n-1)(n+\sigma+\tau), \qquad n \in \mathbb{N}, \tag{2.4.17}$$

and the eigenfunctions are the classical Jacobi orthogonal polynomials

$$P_n^{(\sigma,\tau)}(x) = \begin{cases} 1 & \text{if } n=1, \\ \dfrac{(-1)^{n+1}/w(x)}{\prod\limits_{i=1}^{n-1}(2i-1)} \dfrac{d^{n-1}}{dx^{n-1}}[(1-x^2)^{n-1}w(x)] & \text{if } n>1. \end{cases} \tag{2.4.18}$$

This example illustrates the case where one of the functions $1/f$, q and w is not bounded in the interval (a, b), even though they are all Lebesgue integrable in (a, b).

For $\sigma = \tau = -1/2$, a search on the interval $[-1, 10]$ yields the following approximations of the first four eigenvalues:

$$0., \quad 0.99999, \quad 3.99998, \quad 8.99998, \tag{2.4.19}$$

all eigenvalues are simple; while the corresponding eigenfunctions, i.e., $y(x) = 1$ for $\lambda = 0$, $y(x) = x$ for $\lambda = 1$, $y(x) = 2x^2 - 1$ for $\lambda = 4$ and $y(x) = 4x^3 - 3x$ for $\lambda = 9$, and their approximations are plotted in the following graphs (Figure 2.4.5 ∼ Figure 2.4.8).

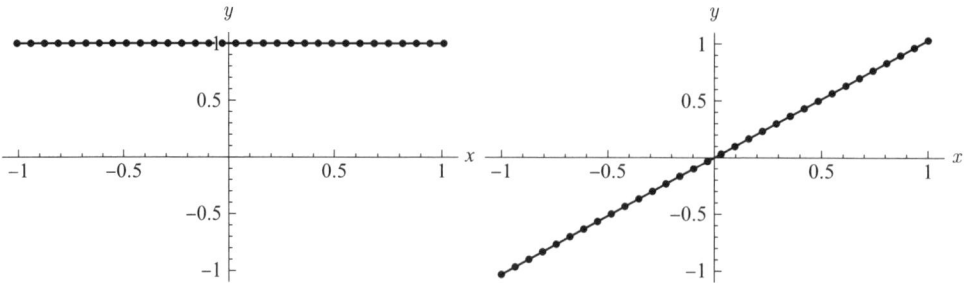

Figure 2.4.5 The eigenfunction for $\lambda = 0$ and its approximation

Figure 2.4.6 The eigenfunction for $\lambda = 1$ and its approximation

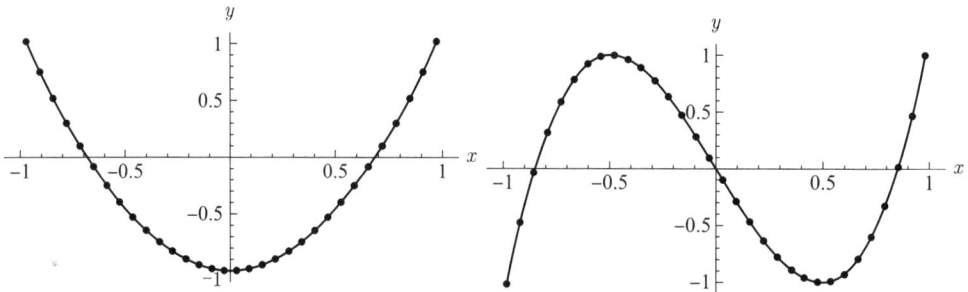

Figure 2.4.7 The eigenfunction for $\lambda = 4$ and its approximation

Figure 2.4.8 The eigenfunction for $\lambda = 9$ and its approximation

In the following example, we will compare the numerical results using codes.

Example 2.4.9 Consider the S-L problem

$$-y'' = \lambda y \text{ in } (-\pi, \pi), \qquad Y(\pi) = e^{i\pi/4}\begin{pmatrix} 2 & 1 \\ 1 & 1 \end{pmatrix} Y(-\pi) \qquad (2.4.20)$$

associated with the Fourier equation. The BC here is non-real, coupled and self-adjoint.

Simple calculations yield that (up to a non-zero constant factor)

$$\Delta(\lambda) = \sqrt{2} - 3\cosh(2\pi\sqrt{-\lambda}) + \frac{\sinh(2\pi\sqrt{-\lambda})}{\sqrt{-\lambda}} - \sqrt{-\lambda}\sinh(2\pi\sqrt{-\lambda}), \quad (2.4.21)$$

whose graph is given below (Figure 2.4.9).

Codes with accuracy = 0.0001 and SLEIGN2 in double precision yield the following approximations of the eigenvalue on the interval $[-5, 5]$:

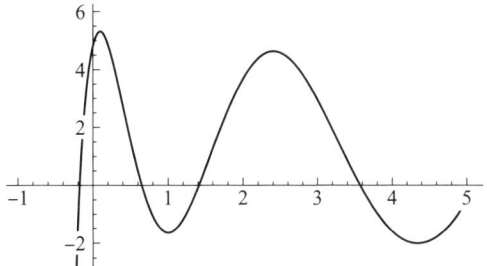

Figure 2.4.9 The graph of the characteristic function

EIGNEIR − RSLP	SLEIGN2[42]	
−0.17244	− 0.09605	
0.64653	0.39208	(2.4.22)
1.41850	1.84674	
3.56577	3.01469	

Using the explicit expression of (2.4.21) for $\Delta(\lambda)$, these eigenvalues can even be approximated on any graphing calculator.

2.5 Examples with λ-dependent BCs

Using Corollary 2.3.1, from the above examples, we obtain the following examples, in which the BCs are all λ-dependent. Again, unless otherwise stated, the parameter gaccuracy in codes takes the default value 10^{-3}; and all S-L problems are treated as general regular problems. Moreover, we will always use S-L problems with an exact eigenvalue 1.

Example 2.5.1 Let $q(x) = 1 + 2/(x^2 + 1)$. Then, 1 is an eigenvalue of the S-L problem

$$-y'' + qy = \lambda y \text{ in } (0, 1), \quad y'(0) = 0 = (2 - \lambda)y(1) - y'(1) \quad (2.5.1)$$

with an eigenfunction $y(x) = x^2 + 1$. In this example, the BC is separated and non-degenerate, and its coefficients of powers of λ are real.

The codes with different values of accuracy yield the following approximations of the simple eigenvalue 1 on the square $[-2, 2] \times [-2, 2]$:

accuracy	approximation	
0.001	1.000457	
0.0001	1.000051	(2.5.2)
0.00001	1.000006	
0.000001	0.999998	

The above results indicate that the codes can obtain desired approximations for different given accuracies, assuming that the ODE solver in Mathematica is sufficiently effective. The graphs of the eigenfunction $y(x) = x^2 + 1$ and its approximation are similar to those in Figure 2.4.1, and hence are omitted.

Example 2.5.2 Let $q(x) = 1 + 2/(x^2 + 1)$. Then, 1 is an eigenvalue of the S-L problem

$$-y'' + qy = \lambda y \text{ in } (0, 1), \qquad (\lambda + 1)y'(0) = 0 = \lambda y(1) - \lambda y'(1) \qquad (2.5.3)$$

with an eigenfunction $y(x) = x^2 + 1$. In this example, the BC is separated, it is degenerate when $\lambda = -1$ or $\lambda = 0$, and its coefficients of powers of λ are real.

A search on the square $[-2, 2] \times [-2, 2]$ yields the simple eigenvalues

$$-1.00005, \quad 0, \quad 1.00005. \qquad (2.5.4)$$

Note that for usual S-L problems, there is no eigenvalue for the two fake BCs; but for an S-L problem with a λ-dependent BC, by Corollary 2.3.2, each value of λ that causes the BC to degenerate is an eigenvalue. So, -1 and 0 are eigenvalues of the S-L problem (2.5.3), and the above numerical results agree with Corollary 2.3.2.

If we change the BC to

$$(\lambda^2 - 1)y'(0) = 0 = \lambda^3 y(1) - \lambda^3 y'(1), \qquad (2.5.5)$$

then it is degenerate when $\lambda = \pm 1$ or $\lambda = 0$, and its coefficients of powers of λ are real.

A search on the square $[-2, 2] \times [-2, 2]$ yields the following eigenvalues, with their multiplicities in front of them:

$$1, -1.00005, \quad 3, 0, \quad 2, 1.00005. \qquad (2.5.6)$$

In the above two examples, if we remove all common factors with zeros from the rows of the coefficient matrices of the BCs, then the corresponding extra eigenvalues are gone. So, we will assume that the BCs are non-degenerate.

If the BC chosen is

$$y'(0) = 0 = [1 + i(1 - \lambda)]y(1) - y'(1), \tag{2.5.7}$$

then it is separated and non-degenerate, and its coefficients of powers of λ are non-real.

A search on the square $[-2, 30] \times [-2, 2]$ yields the simple eigenvalue

$$1.00001, \quad 5.5012 + 1.1969i, \quad 24.9052 + 1.9009i. \tag{2.5.8}$$

We take

$$\begin{pmatrix} 2 - \lambda & 0 & -1/2 & 0 \\ 0 & -1/2 & -1 & 2 - \lambda \end{pmatrix} \begin{pmatrix} \boldsymbol{Y}(0) \\ \boldsymbol{Y}(1) \end{pmatrix} = \boldsymbol{0}, \tag{2.5.9}$$

then the BC is coupled and non-degenerate, and its coefficients of powers of λ are real.

A search on the square $[-2, 30] \times [-2, 2]$ yields the simple eigenvalues

$$1.00005, \quad 1.79823, \quad 6.0899, \quad 24.4424. \tag{2.5.10}$$

If we pick

$$\begin{pmatrix} 1 + i(1 - \lambda) & 0 \\ 0 & 1 \end{pmatrix} \boldsymbol{Y}(0) + \begin{pmatrix} -1/2 & 0 \\ -1 & 1 + i(1 - \lambda) \end{pmatrix} \boldsymbol{Y}(1) = \boldsymbol{0}, \tag{2.5.11}$$

then the BC is coupled and non-degenerate, and its coefficients of powers of λ are non-real.

A search on the square $[-2, 20] \times [-3, 3]$ finds the simple eigenvalues

$$1.00005, \quad 1.01001 - 1.56155i, \quad 4.8071 - 0.1345i. \tag{2.5.12}$$

Example 2.5.3 Let $k \in \mathbb{C}$, and define $q(x) = k^2 \cos^2 x - 3k \sin x$. Then, 1 is an eigenvalue of the S-L problem

$$-y'' + qy = \lambda y \text{ in } (0, 2\pi), \quad \boldsymbol{A}(\lambda)\boldsymbol{Y}(0) + \boldsymbol{B}(\lambda)\boldsymbol{Y}(2\pi) = \boldsymbol{0} \tag{2.5.13}$$

with an eigenfunction $y(x) = e^{k \sin x} \cos x$, where

$$\boldsymbol{A}(\lambda) = \begin{pmatrix} 2 - \lambda & 0 \\ 0 & 1 \end{pmatrix}, \quad \boldsymbol{B}(\lambda) = \begin{pmatrix} -1 & 0 \\ 0 & -2 + \lambda \end{pmatrix}. \tag{2.5.14}$$

In this example, the BC is coupled and non-degenerate, and its coefficients of powers of λ are real.

For several values of k, a search on the square $[-2, 2] \times [-2, 2]$ yields the following eigenvalues, with their multiplicities in front of them:

k	multiplicities and approximations
-2	1, 1.00005, 1, 1.81172 + 0.91897i, 1, 1.81172 - 0.91897i

−1	1, −1.65422, 1, 1.00005, 1, 1.81172 + 0.91897i,
	1, 1.81172 − 0.91897i
0	1, 0.01036, 2, 1.00005
1	1, −1.48271, 1, 1.00005
2	1, 1.00005

(2.5.15)

Example 2.5.4 For $\sigma, \tau \in \mathbb{R}$, let
$$f(x) = (1-x)^{\sigma+1}(1+x)^{\tau+1}, \qquad w(x) = (1-x)^{\sigma}(1+x)^{\tau}. \tag{2.5.16}$$

By Corollary 2.3.1, when $\sigma, \tau \in (-1, 0)$, 1 is an eigenvalue of the S-L problem
$$\begin{cases} -(fy')' = \lambda wy \text{ in } (-1, 1), \\ (\lambda - 1)y(-1) + \lambda(fy')(-1) = 0 = (\lambda - 1)y(1) + \lambda(fy')(1) \end{cases} \tag{2.5.17}$$

with an eigenfunction $y(x) = x$. In this example, the BC is separated and non-degenerate, and its coefficients of powers of λ are real.

For $\sigma = \tau = -1/2$, a search on the square $[-2, 2] \times [-2, 2]$ yields the simple eigenvalues:

$$0.57396 - 0.36898i, \quad 0.57396 + 0.36898i, \quad 1.00005. \tag{2.5.18}$$

This example illustrates the case where one of the functions $1/f$, q and w is not bounded in the interval (a, b), even though they are all Lebesgue integrable in (a, b).

2.6 Oscillations of eigenfunctions for discontinuous Sturm-Liouville problems

Consider a class of discontinuous eigenvalue problems consisting of the S-L equation
$$-(fy')' + qy = \lambda wy, \quad x \in (a, c) \cup (c, b), \tag{2.6.1}$$

the boundary conditions
$$y(a) \cos \alpha + (fy')(a) \sin \alpha = 0, \tag{2.6.2}$$
$$y(b) \cos \beta + (fy')(b) \sin \beta = 0, \tag{2.6.3}$$

and the transmission conditions
$$y(c^+) - \delta y(c^-) = 0, \tag{2.6.4}$$

$$(fy')(c^+) - \delta_1 y(c^-) - \delta_2 (fy')(c^-) = 0, \tag{2.6.5}$$

where $0 \leqslant \alpha < \pi, 0 < \beta \leqslant \pi; \delta, \delta_1, \delta_2 \in \mathbb{R}$ and $\delta \delta_2 > 0$, $y(c^+) = \lim_{x \to c+0} y(x)$, $y(c^-) = \lim_{x \to c-0} y(x)$,

$$-\infty < a < c < b < +\infty, \quad q \in L((a,c) \cup (c,b), \mathbb{R}), f, w > 0, \tag{2.6.6}$$

$L((a,c) \cup (c,b), \mathbb{R})$ denotes the space of real-valued Lebesgue integrable functions in $(a,c) \cup (c,b)$, $\lambda \in \mathbb{C}$ is the spectral parameter.

2.6.1 Basic conclusions

For every $\lambda \in \mathbb{C}$, let $\xi_{1,-}(x, \lambda)$ and $\xi_{2,-}(x, \lambda)$ be the solutions of (2.6.1) restricted to (a,c) satisfying

$$\xi_{1,-}(a) = 1, (f\xi'_{1,-})(a) = 0; \tag{2.6.7}$$

$$\xi_{2,-}(a) = 0, (f\xi'_{2,-})(a) = 1 \tag{2.6.8}$$

respectively. Then

$$\xi_{1,+}(c^+) = \delta \xi_{1,-}(c^-), \ (f\xi'_{1,+})(c^+) = \delta_1 \xi_{1,-}(c^-) + \delta_2 (f\xi'_{1,-})(c^-); \tag{2.6.9}$$

$$\xi_{2,+}(c^+) = \delta \xi_{2,-}(c^-), \ (f\xi'_{2,+})(c^+) = \delta_1 \xi_{2,-}(c^-) + \delta_2 (f\xi'_{2,-})(c^-). \tag{2.6.10}$$

Let $\zeta_{1,+}(x, \lambda)$ and $\zeta_{2,+}(x, \lambda)$ be the solutions to (2.6.1) restricted to (c,b) satisfying

$$\zeta_{1,+}(c^+) = 1, (f\zeta'_{1,+})(c^+) = 0; \tag{2.6.11}$$

$$\zeta_{2,+}(c^+) = 0, (f\zeta'_{2,+})(c^+) = 1 \tag{2.6.12}$$

respectively. Then

$$\zeta_{1,-}(c^-) = \frac{1}{\delta}\zeta_{1,+}(c^+), (f\zeta'_{1,-})(c^-) = \frac{1}{\delta, \delta_2}\left(-\delta_1\zeta_{1,+}(c^+) + \delta(f\zeta'_{1,+})(c^+)\right); \tag{2.6.13}$$

$$\zeta_{2,-}(c^-) = \frac{1}{\delta}\zeta_{2,+}(c^+), (f\zeta'_{2,-})(c^-) = \frac{1}{\delta, \delta_2}\left(-\delta_1\zeta_{2,+}(c^+) + \delta(f\zeta'_{2,+})(c^+)\right). \tag{2.6.14}$$

We use

$$\xi_1(x, \lambda) = \begin{cases} \xi_{1,-}(x, \lambda), x \in (a, c), \\ \xi_{1,+}(x, \lambda), x \in (c, b) \end{cases} \tag{2.6.15}$$

to denote the solutions to (2.6.1) in the interval $(a, c) \cup (c, b)$ satisfying the initial conditions in (2.6.7) and (2.6.9). The definitions of $\xi_2(x, \lambda)$, $\zeta_1(x, \lambda)$ and $\zeta_2(x, \lambda)$ are similar to $\xi_1(x, \lambda)$. For each $x \in (a, c) \cup (c, b)$, $\xi_{i,\pm}(x, \lambda), \zeta_{i,\pm}(x, \lambda)$ $(i = 1, 2)$ are entire functions of λ. A number $\lambda \in \mathbb{R}$ is an eigenvalue of the S-L problems (2.6.1)∼(2.6.6) if and

only if
$$\sum_{i,j=1}^{2} D_{i,j}\Psi_{i,j} = 0, \tag{2.6.16}$$

where
$$D = \begin{pmatrix} -\sin\alpha\cos\beta & \cos\alpha\cos\beta \\ -\sin\alpha\sin\beta & \cos\alpha\sin\beta \end{pmatrix}, \tag{2.6.17}$$

$$\Psi = \begin{pmatrix} \zeta_{1,+}(b,\lambda) & \zeta_{2,+}(b,\lambda) \\ (f\zeta'_{1,+})(b,\lambda) & (f\zeta'_{2,+})(b,\lambda) \end{pmatrix} \begin{pmatrix} \xi_{1,+}(c^+,\lambda) & \xi_{2,+}(c^+,\lambda) \\ (f\xi'_{1,+})(c^+,\lambda) & (f\xi'_{2,+})(c^+,\lambda) \end{pmatrix}. \tag{2.6.18}$$

The S-L problems (2.6.1)~(2.6.6) have infinitely, but countably many eigenvalues, the eigenvalues are real, simple and boundary from below. They can be indexed using the natural numbers so that

$$\lambda_1 < \lambda_2 < \lambda_3 < \cdots \to +\infty. \tag{2.6.19}$$

Let y be a non-trivial real solution to (2.6.1) with $\lambda \in \mathbb{R}$. Then, there are two unique absolutely continuous functions ρ and θ in $(a,c) \cup (c,b)$ such that

$$\rho(x,\lambda) \neq 0 \text{ for } x \in (a,c) \cup (c,b), \tag{2.6.20}$$

$$y = \rho\sin\theta, \quad fy' = \rho\cos\theta, \quad 0 \leqslant \theta(a,\lambda) < \pi, \tag{2.6.21}$$

where
$$\rho(x,\lambda) = \begin{cases} \rho_-(x,\lambda), & x \in (a,c), \\ \rho_+(x,\lambda), & x \in (c,b). \end{cases} \quad \theta(x,\lambda) = \begin{cases} \theta_-(x,\lambda), & x \in (a,c), \\ \theta_+(x,\lambda), & x \in (c,b), \end{cases} \tag{2.6.22}$$

and either
$$\tan[\theta_+(c^+)] = \frac{\delta\sin[\theta_-(c^-)]}{\delta_1\sin[\theta_-(c^-)] + \delta_2\cos[\theta_-(c^-)]}, \tag{2.6.23}$$

or
$$\cot[\theta_+(c^+)] = \frac{\delta_1\sin[\theta_-(c^-)] + \delta_2\cos[\theta_-(c^-)]}{\delta\sin[\theta_-(c^-)]}. \tag{2.6.24}$$

Actually,
$$\rho = \pm\sqrt{(y^2) + (fy')^2} \tag{2.6.25}$$

and
$$\text{either } \tan\theta = \frac{y}{fy'} \quad \text{or} \quad \cot\theta = \frac{fy'}{y} \tag{2.6.26}$$

in each compact subinterval of (a, c) and (c, b). The function θ is usually called the Prüfer angle of the solution y. From (2.6.1) and (2.6.25) we obtain that

$$\rho' = \rho \left(\frac{1}{f} + q - \lambda w \right) \cos\theta \sin\theta \qquad (2.6.27)$$

in each compact subintervals of (a, c) and (c, b), which is equivalent to

$$\begin{cases} \rho_1(x, \lambda) = \rho_1(a, \lambda)\exp\int_a^x \left(\frac{1}{f_1} + q_1 - \lambda w_1 \right) \cos\theta \sin\theta \mathrm{d}\tau, & x \in [a, c], \\ \rho_2(x, \lambda) = \rho_2(c, \lambda)\exp\int_c^x \left(\frac{1}{f_2} + q_2 - \lambda w_2 \right) \cos\theta \sin\theta \mathrm{d}\tau, & x \in [c, b]. \end{cases} \qquad (2.6.28)$$

From (2.6.1) and (2.6.26) we deduce that

$$\theta' = \frac{1}{f}\cos^2\theta + (\lambda w - q)\sin^2\theta \qquad (2.6.29)$$

in each compact subintervals of (a, c) and (c, b). For each $n \in \mathbb{N}$, there is a unique eigenvalue, to be denoted by λ_n, of the problem such that (2.6.1) with $\lambda = \lambda_n$ has real solutions whose Prüfer angle $\theta(\cdot, \lambda_n)$ satisfies

$$\theta(a, \lambda_n) = \alpha, \quad \theta(b, \lambda_n) = \beta + (n-1)\pi. \qquad (2.6.30)$$

2.6.2 Main results

The following result is a consequence of Sturm's comparison theorem.

Theorem 2.6.1 Assume that $g_{i,-}(x)$, $h_{i,-}(x) \in L((a, c), \mathbb{R})$, and $g_{i,+}(x)$, $h_{i,+}(x) \in L((c, b), \mathbb{R})$, for $i=1$ and 2, satisfy

$$f_1 \leqslant f_2, g_1 \leqslant g_2 \text{ a.e. in } (a, c) \cup (c, b). \qquad (2.6.31)$$

Let θ_i be a solution to

$$\theta' = f_i \cos^2\theta + g_i \sin^2\theta \text{ in } (a, c) \cup (c, b), \qquad (2.6.32)$$

where

$$g_i(x) = \begin{cases} g_{i,-}(x), & x \in (a, c), \\ g_{i,+}(x), & x \in (c, b), \end{cases} \quad f_i(x) = \begin{cases} f_{i,-}(x), & x \in (a, c), \\ f_{i,+}(x), & x \in (c, b), \end{cases} \qquad (2.6.33)$$

$$\theta_i(x) = \begin{cases} \theta_{i,-}(x), & x \in (a, c), \\ \theta_{i,+}(x), & x \in (c, b), \end{cases} \qquad (2.6.34)$$

and either

$$\tan(\theta_{i,+}(c^+)) = \frac{\delta \sin\gamma}{\delta_1 \sin\gamma + \delta_2 \cos\gamma} \text{ or } \cot(\theta_{i,+}(c^+)) = \frac{\delta_1 \sin\gamma + \delta_2 \cos\gamma}{\delta \sin\gamma}, \ \gamma = \theta_{i,-}(c^-). \tag{2.6.35}$$

Then we have that

(1) If $\theta_1(a) < \theta_2(a)$, then $\theta_1 < \theta_2$ on $[a,c) \cup (c,b]$.

(2) If $\theta_1(x_*) = \theta_2(x_*)$ for some $x_* \in [a,c)$, then $\theta_1 > \theta_2$ on $[a,x_*)$, and $\theta_1 < \theta_2$ on $[x_*,c) \cup (c,b]$. If $\theta_1(x_*) = \theta_2(x_*)$ for some $x_* \in (c,b]$, then $\theta_1 < \theta_2$ on $(x_*,b]$, and $\theta_1 > \theta_2$ on $[a,c) \cup (c,x_*]$.

(3) If $\theta_1(b) = \theta_2(b)$ and either $f_1 \neq 0$ or $f_2 \neq 0$, $g_1 < g_2$ a.e. on $[x_*,b]$ (or in each compact subintervals of $[x_*,c) \cup (c,b]$) for some $x_* \in [a,c) \cup (c,b)$, then $\theta_1 > \theta_2$ on $[a,c) \cup (c,b)$.

(4) If $\theta_1(a) = \theta_2(a)$ and either $f_1 \neq 0$ or $f_2 \neq 0$, $g_1 < g_2$ a.e. on $[a,x_*]$ (or in each compact subintervals of $[a,c) \cup (c,x_*]$) for some $x_* \in (a,c) \cup (c,b]$, then $\theta_1 < \theta_2$ on $(a,c) \cup (c,b]$.

(5) If $\theta_1(b) > \theta_2(b)$, then $\theta_1 > \theta_2$ on $[a,c) \cup (c,b]$.

Proof. The proof is similar to [3]. □

The following result is a direct consequence of Theorem 2.6.1.

Corollary 2.6.1 Let y be a non-trivial real solution to the S-L equation (2.6.1) such that $y(a,\lambda)$ and $(fy')(a,\lambda)$ are independent of λ. Then, on each compact subinterval of $[a,c)$ and $(c,b]$, the Prüfer angle $\theta(x,\lambda)$ of y is strictly increasing in λ on $\mathbb{R}$.

In particular, if furthermore $p > 0$ a.e. in $(a,c) \cup (c,b)$ and either

$$y(a,\lambda) > 0 \text{ or } y(a,\lambda) = 0 \text{ and } (fy')(a,\lambda) > 0, \tag{2.6.36}$$

then for any $\lambda_* < \lambda^*$,

$$y(x,\lambda_*) > y(x,\lambda^*) \quad \text{for } x \in (a,x^*] \text{ or } x \in (a,c) \cup (c,x^*], \tag{2.6.37}$$

where x^* is the first zero of $y(\cdot,\lambda^*)$ in $(a,c) \cup (c,b]$.

Proof. The strict increase is a direct consequence of Theorem 2.6.1, since $\theta(a,\lambda)$ is independent of λ in this case.

To see the particular part, we already have that

$$\theta(x,\lambda_*) < \theta(x,\lambda^*), x \in (a,c) \cup (c,b]. \tag{2.6.38}$$

Note that (2.6.36) implies that $\exists \varepsilon > 0$ s.t. $y(\cdot,\lambda_*) > 0$ and $y(\cdot,\lambda^*) > 0$ in $(0,\varepsilon)$. Let x_* be the first zero of $y(\cdot,\lambda_*)$ in $(a,c) \cup (c,b]$ and set $x_\# = \min\{x_*,x^*\}$. Then, (2.6.38) yields that

$$\frac{(fy')(x,\lambda_*)}{y(x,\lambda_*)} > \frac{(fy')(x,\lambda^*)}{y(x,\lambda^*)}, \quad x \in (a, x_\#). \tag{2.6.39}$$

The assumption that $f > 0$ a.e. in $(a, c) \cup (c, b)$ and (2.6.39) imply that a.e. in $(a, x_\#)$,

$$\frac{y'(x,\lambda_*)}{y(x,\lambda_*)} > \frac{y'(x,\lambda^*)}{y(x,\lambda^*)}, \text{i.e.,} \ln\left(\frac{y(x,\lambda_*)}{y(x,\lambda^*)}\right)' > 0, \tag{2.6.40}$$

which together with the independence of $y(a, \lambda)$ and $(fy')(a, \lambda)$ on λ yield that

$$y(\cdot, \lambda_*) > y(\cdot, \lambda^*) \text{ on } (a, t_\#]. \tag{2.6.41}$$

Here for the case where $y(a, \lambda) = 0$, we use the fact that

$$\lim_{x \to a^+} \frac{\phi_{12}(x, \lambda_*)}{\phi_{12}(x, \lambda^*)} \to 1. \tag{2.6.42}$$

Therefore, $x_* > x^*$ and (2.6.37) is true. $\square$

Lemma 2.6.1 Let $\varrho(\lambda) = \varrho_-(\lambda) + \varrho_+(\lambda)$ denotes the number of zeros of $\xi(x, \lambda)$ in the interval $(a, c) \cup (c, b)$. $M = \max\limits_{x \in (a, c) \cup (c, b)} |q(x)|$. Then

(1) $\exists M > 0$, s.t. when $\lambda < -M$, $\varrho(\lambda) = 0$.
(2) $\lim\limits_{\lambda \to +\infty} \varrho(\lambda) = +\infty$.

Proof. (1) When $\lambda + M < 0$, the solution to the initial value problem

$$\begin{cases} y'' + (\lambda + M)y = 0, & x \in (a, c), \\ y(a, \lambda) = \sin\alpha, \ y'(a, \lambda) = -\cos\alpha \end{cases} \tag{2.6.43}$$

is as follows:

$$\xi_-(x, \lambda) = \frac{\sqrt{-(\lambda + M)}\sin\alpha - \cos\alpha}{2\sqrt{-(\lambda + M)}} e^{\sqrt{-(\lambda+M)}(x-a)} +$$

$$\frac{\sqrt{-(\lambda + M)}\sin\alpha + \cos\alpha}{2\sqrt{-(\lambda + M)}} e^{-\sqrt{-(\lambda+M)}(x-a)} \tag{2.6.44}$$

$$= \left(\frac{1}{2}\sin\alpha - \frac{\cos\alpha}{2\sqrt{-(\lambda + M)}}\right) e^{\sqrt{-(\lambda+M)}(x-a)} +$$

$$\left(\frac{1}{2}\sin\alpha + \frac{\alpha}{2\sqrt{-(\lambda + M)}}\right) e^{-\sqrt{-(\lambda+M)}(x-a)}.$$

From (2.6.44), when $\lambda \to -\infty$, $\xi_-(x, \lambda) > 0$. So $\varrho_-(\lambda) = 0$.
Then, the solution to the initial value problem

Chapter 2 Approximations of eigenvalues and eigenfunctions

$$\begin{cases} y'' + (\lambda + M)y = 0, & x \in (c, b), \\ y(c^+, \lambda) = \gamma_1 y(c^-, \lambda), \\ y'(c^+, \lambda) = \gamma_3 y(c^-, \lambda) + \gamma_4 y'(c^-, \lambda) \end{cases} \qquad (2.6.45)$$

is as follows:

$$\begin{aligned}\xi_+(x, \lambda) &= \frac{\sqrt{-(\lambda + M)}\,\gamma_1 \xi(c^-, \lambda)}{2\sqrt{-(\lambda + M)}} e^{\sqrt{-(\lambda+M)}\,(x-c^+)} + \\ &\quad \frac{\gamma_3 \xi(c^-, \lambda) + \gamma_4 \xi'(c^-, \lambda)}{2\sqrt{-(\lambda + M)}} e^{\sqrt{-(\lambda+M)}\,(x-c^+)} + \\ &\quad \frac{\sqrt{-(\lambda + M)}\,\gamma_1 \xi(c^-, \lambda)}{2\sqrt{-(\lambda + M)}} e^{-\sqrt{-(\lambda+M)}\,(x-c^+)} - \\ &\quad \frac{\gamma_3 \xi(c^-, \lambda) + \gamma_4 \xi'(c^-, \lambda)}{2\sqrt{-(\lambda + M)}} e^{-\sqrt{-(\lambda+M)}\,(x-c^+)} \end{aligned} \qquad (2.6.46)$$

$$\begin{aligned}&= \frac{1}{2}\Big[\gamma_1 \xi(c^-, \lambda) + \frac{\gamma_3 \xi(c^-, \lambda) + \gamma_4 \xi'(c^-, \lambda)}{\sqrt{-(\lambda + M)}}\Big] e^{\sqrt{-(\lambda+M)}\,(x-c^+)} + \\ &\quad \frac{1}{2}\Big[\gamma_1 \xi(c^-, \lambda) - \frac{\gamma_3 \xi(c^-, \lambda) + \gamma_4 \xi'(c^-, \lambda)}{\sqrt{-(\lambda + M)}}\Big] e^{-\sqrt{-(\lambda+M)}\,(x-c^+)}. \end{aligned} \qquad (2.6.47)$$

From (2.6.47), when $\lambda \to -\infty$, $\xi_+(x, \lambda) > 0$, $\varrho_+(\lambda) = 0$. So

$$\exists M > 0, \quad \forall \lambda < -M, \quad \varrho(\lambda) = 0. \qquad (2.6.48)$$

(2) When $\lambda - M > 0$, the solution to the initial value problem

$$\begin{cases} y'' + (\lambda - M)y = 0, & x \in (a, c), \\ y(a, \lambda) = \sin \alpha, \quad y'(a, \lambda) = -\cos \alpha \end{cases} \qquad (2.6.49)$$

is as follows:

$$\begin{aligned}\xi_-(x, \lambda) &= \sin \alpha \cos \sqrt{\lambda - M}\,(x - a) - \frac{\cos \alpha}{\sqrt{\lambda - M}} \sin \sqrt{\lambda - M}\,(x - a) \\ &= \sqrt{\sin^2 \alpha + \frac{\cos^2 \alpha}{\lambda - M}} \Bigg[\frac{\sin \alpha}{\sqrt{\sin^2 \alpha + \frac{\cos^2 \alpha}{\lambda - M}}} \cos \sqrt{\lambda - M}\,(x - a) - \\ &\quad \frac{\cos \alpha}{\sqrt{(\lambda - M)\left(\sin^2 \alpha + \frac{\cos^2 \alpha}{\lambda - M}\right)}} \sin \sqrt{\lambda - M}\,(x - a) \Bigg] \\ &\doteq B \sin(\sqrt{\lambda - M}\,x + \theta). \end{aligned} \qquad (2.6.50)$$

The number of zeros of (2.6.50) in the interval (a, c) is

$$\frac{c^- - a}{\pi}\sqrt{\lambda - M} \text{ or } \frac{c^- - a}{\pi}\sqrt{\lambda - M} + 1.$$

Then, the solution to the initial value problem

$$\begin{cases} y'' + (\lambda - M)y = 0, & x \in (c, b), \\ y(c^+, \lambda) = \gamma_1 y(c^-, \lambda), \\ y'(c^+, \lambda) = \gamma_3 y(c^-, \lambda) + \gamma_4 y'(c^-, \lambda) \end{cases} \quad (2.6.51)$$

is as follows:

$$\xi_+(x, \lambda) = \gamma_1 \xi(c^-, \lambda) \cos\sqrt{\lambda - M}(x - c^+) -$$
$$\frac{\gamma_3, \xi(c^-, \lambda) + \gamma_4, \xi'(c^-, \lambda)}{\sqrt{\lambda - M}} \sin\sqrt{\lambda - M}(x - c^+) \quad (2.6.52)$$
$$\doteq C \sin(\sqrt{\lambda - M}x + \delta).$$

The number of zeros of (2.6.52) in the interval (c, b) is

$$\frac{b - c^+}{\pi}\sqrt{\lambda - M} \text{ or } \frac{b - c^+}{\pi}\sqrt{\lambda - M} + 1. \quad (2.6.53)$$

Therefore, the number of zeros of $\phi(x, \lambda)$ in $(a, c) \cup (c, b)$ is at least

$$\frac{c^- - a}{\pi}\sqrt{\lambda - M} + \frac{b - c^+}{\pi}\sqrt{\lambda - M}. \quad (2.6.54)$$

Hence, when $\lambda > M$, the number of zeros of $\zeta(x, \lambda)$ in $(a, c) \cup (c, b)$ is no less than $\frac{c^- - a}{\pi}\sqrt{\lambda - M} + \frac{b - c^+}{\pi}\sqrt{\lambda - M}$, that is,

$$\lim_{\lambda \to +\infty} \varrho(\lambda) = +\infty. \quad (2.6.55)$$

Combining (2.6.48) and (2.6.55), the conclusion is confirmed. □

Theorem 2.6.2 For each $n \in \mathbb{N}$, any eigenfunction for λ_n ($n = 1, 2, 3, \cdots$) of the S-L equations (2.6.1)$\sim$(2.6.6) have exactly $n - 1$ zeros in the open interval $(a, c) \cup (c, b)$.

Proof. See [27] for the proof. □

2.6.3 Numerical examples

In this section, numerical examples are used to illustrate the above conclusions.

Example 2.6.1 Let

$$q(x) = \begin{cases} 1 + 2/(1 + x^2), \\ 1 + 2/(2 + x^2). \end{cases} \quad (2.6.56)$$

Consider the S-L problem

$$\begin{cases} -y'' + qy = \lambda y \text{ on } [0,1) \cup (1,2], \\ y'(0) = 0 = 2y(2) - 3y'(2), \\ y(1^+) - 1.5y(1^-) = 0, \\ 3y'(1^+) - y(1^-) - 2y'(1^-) = 0. \end{cases} \quad (2.6.57)$$

Some eigenvalues and eigenfunctions are shown in Figure 2.6.1~ Figure 2.6.6.

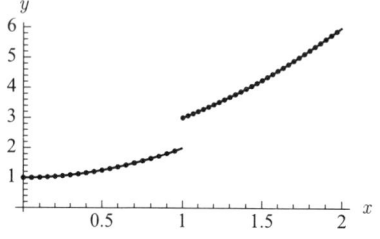

Figure 2.6.1 $\lambda_1 = 1.0$

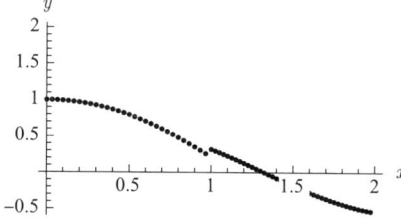

Figure 2.6.2 $\lambda_2 = 4.6217$

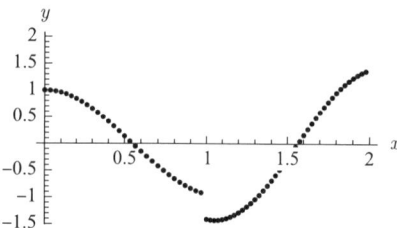

Figure 2.6.3 $\lambda_3 = 11.0982$

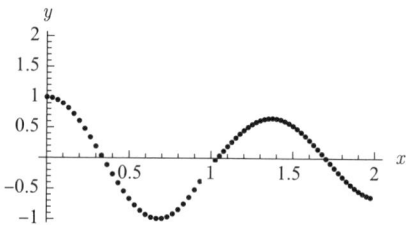

Figure 2.6.4 $\lambda_4 = 24.0541$

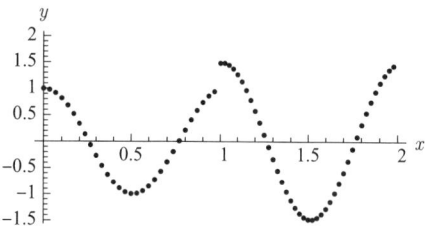

Figure 2.6.5 $\lambda_5 = 40.6844$

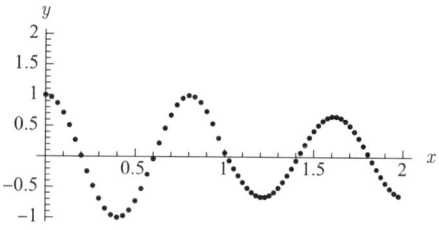

Figure 2.6.6 $\lambda_6 = 63.5179$

Chapter 3 Computing the indices of eigenvalues

In this chapter, we consider a self-adjoint S-L problem, i.e., the spectral problem consisting of an S-L equation

$$-(fy')' + qy = \lambda w y \quad \text{in } (a,b) \tag{3.0.1}$$

and a self-adjoint BC, where

$$-\infty \leqslant a < b \leqslant +\infty, \quad 1/f, q, w \in L((a,b), \mathbb{R}), \quad w > 0 \text{ a.e. in } (a,b), \tag{3.0.2}$$

and $\lambda \in \mathbb{C}$ is the so-called spectral parameter. Note that the leading coefficient function f is allowed to change the sign in (a,b), i.e., is indefinite: both $\{x \in (a,b); f(x) > 0\}$ and $\{x \in (a,b); f(x) < 0\}$ have a positive Lebesgue measure.

Throughout this chapter, we fix the S-L equation (3.0.1) satisfying (3.0.2), and all BCs considered are self-adjoint.

3.1 Notation and theoretic results

By the integrability conditions in (3.0.2), all solutions y to (3.0.1) and their quasi-derivatives fy' have finite limits at both a and b, even when $a = -\infty$ or $b = +\infty$. If $a = -\infty$ or $b = +\infty$, then the notation $[a,b]$ still has an evident meaning, and the interval is given an obvious topology to make it compact. For every $\lambda \in \mathbb{C}$, let $\phi_{11}(\cdot, \lambda)$ and $\phi_{12}(\cdot, \lambda)$ be the solutions to (3.0.1) determined by the initial conditions

$$\phi_{11}(a, \lambda) = 1, \ (f\phi'_{11})(a, \lambda) = 0, \quad \phi_{12}(a, \lambda) = 0, \ (f\phi'_{12})(a, \lambda) = 1. \tag{3.1.1}$$

Then, they form a fundamental set of solutions to (3.0.1). We denote $f\phi'_{11}$ and $f\phi'_{12}$ by ϕ_{21} and ϕ_{22}, respectively. Set

$$\Phi(x, \lambda) = \begin{pmatrix} \phi_{11}(x, \lambda) & \phi_{12}(x, \lambda) \\ \phi_{21}(x, \lambda) & \phi_{22}(x, \lambda) \end{pmatrix}, \quad x \in [a,b], \ \lambda \in \mathbb{C}. \tag{3.1.2}$$

Then, $\Phi(x, \lambda)$ satisfies the matrix form of (3.0.1), i.e.,

$$\Phi'(x, \lambda) = \begin{pmatrix} 0 & 1/f(x) \\ q(x) - \lambda w(x) & 0 \end{pmatrix} \Phi(x, \lambda) \quad \text{in } (a,b), \tag{3.1.3}$$

and the initial condition $\boldsymbol{\Phi}(a, \lambda) = \boldsymbol{I}$. We call $\boldsymbol{\Phi}$ the fundamental solution matrix to (3.0.1). Note that $\boldsymbol{\Phi}(x, \lambda) \in \mathrm{SL}(2,\mathbb{R})$ for $x \in [a, b]$ and $\lambda \in \mathbb{R}$.

For any solution y to (3.0.1), let

$$\boldsymbol{Y}(x) = \begin{pmatrix} y(x) \\ (fy')(x) \end{pmatrix}, \qquad x \in [a, b]. \tag{3.1.4}$$

Then, BCs are specified by algebraic systems of the form

$$\boldsymbol{A}\boldsymbol{Y}(a) + \boldsymbol{B}\boldsymbol{Y}(b) = \boldsymbol{0}, \tag{3.1.5}$$

where $\boldsymbol{A}$ and $\boldsymbol{B}$ are 2×2 complex matrices such that the 2×4 matrix $(\boldsymbol{A} \,|\, \boldsymbol{B})$ has rank 2. Note that equivalent algebraic systems define the same BC. Bold-faced capital Latin letters, such as $\mathbf{A}$, are used to stand for BCs.

Each separated self-adjoint BC has its standard form

$$\cos \alpha \cdot y(a) - \sin \alpha \cdot (fy')(a) = 0 = \cos \beta \cdot y(b) - \sin \beta \cdot (fy')(b) \tag{3.1.6}$$

with $\alpha \in [0, \pi)$ and $\beta \in (0, \pi]$. We denote this BC by $\mathbf{S}_{\alpha,\beta}$. In this notation, the Dirichlet BC is $\mathbf{S}_{0,\pi}$, and the Neumann BC is $\mathbf{S}_{\pi/2,\pi/2}$. Moreover, $\mathbf{S}_{\alpha,\beta}$ can be defined for any $\alpha, \beta \in \mathbb{R}$, and ranges of α and β different from $[0, \pi)$ and $(0, \pi]$ will also be used later.

Every coupled self-adjoint BC can be written in the form

$$\boldsymbol{Y}(b) = \mathrm{e}^{\mathrm{i}\gamma} \boldsymbol{K} \boldsymbol{Y}(a) \tag{3.1.7}$$

with $\gamma \in [0, \pi)$ and $\boldsymbol{K} \in \mathrm{SL}(2,\mathbb{R})$. Sometimes, it is convenient to allow other ranges of γ, such as $(-\pi, \pi)$ and $\mathbb{R}$.

For any non-trivial real solution to (3.0.1), there are two unique absolutely continuous functions ρ and θ on $[a, b]$ such that $\rho(x, \lambda) \neq 0$ for all $x \in [a, b]$, and

$$y = \rho \sin \theta, \quad fy' = \rho \cos \theta, \quad 0 \leqslant \theta(a, \lambda) < \pi. \tag{3.1.8}$$

The function θ is called the Prüfer angle of the solution y. The zeros of y on $[a, b]$ are exactly the points of $[a, b]$ where θ attains an integer multiple of π. Note that y satisfies the self-adjoint BC $\mathbf{S}_{\alpha,\beta}$ if and only if

$$\theta(a, \lambda) = \alpha, \quad \theta(b, \lambda) = \beta + (n-1)\pi \text{ for some } n \in \mathbb{Z}. \tag{3.1.9}$$

If an eigenvalue has geometric multiplicity 1 and a real eigenfunction, then all its real eigenfunctions share the same Prüfer angle. In this case, the Prüfer angle is called the Prüfer angle of the eigenvalue.

In the positive f case, the Prüfer angle characterization of eigenvalues for separated self-adjoint BCs is well known. In the indefinite f case, the indices of the eigenvalues for separated self-adjoint BCs are defined using a similar characterization.

Lemma 3.1.1 Assume that f changes the sign in (a,b), and fix an $\alpha \in [0,\pi)$ and a $\beta \in (0,\pi]$. Then, for each $n \in \mathbb{Z}$, there is a unique eigenvalue for $\mathbf{S}_{\alpha,\beta}$, to be denoted by $\lambda_n = \lambda_n(\mathbf{S}_{\alpha,\beta})$, such that its Prüfer angle θ satisfies

$$\theta(b,\lambda_n) = \beta + (n-1)\pi. \tag{3.1.10}$$

Moreover, $\lambda_n \to -\infty$ as $n \to -\infty$, and $\lambda_n \to +\infty$ as $n \to +\infty$.

Proof. See Theorem 2.2 in [97]. □

Definition 3.1.1 For each $\boldsymbol{K} \in \mathrm{SL}(2,\mathbb{R})$, define $\beta_{0,\boldsymbol{K}}, \beta_{1,\boldsymbol{K}} \in (0,\pi]$ by

$$\begin{aligned} \tan\beta_{0,\boldsymbol{K}} &= k_{12}/k_{22} \quad \text{or} \quad \cot\beta_{0,\boldsymbol{K}} = k_{22}/k_{12}, \\ \tan\beta_{1,\boldsymbol{K}} &= k_{11}/k_{21} \quad \text{or} \quad \cot\beta_{1,\boldsymbol{K}} = k_{21}/k_{11}. \end{aligned} \tag{3.1.11}$$

For a fixed $\lambda_* \in \mathbb{R}$, we set

$$\boldsymbol{\Psi} = \begin{pmatrix} \psi_{11} & \psi_{12} \\ \psi_{21} & \psi_{22} \end{pmatrix} := \boldsymbol{\Phi}(b,\lambda_*), \tag{3.1.12}$$

and call it the transfer matrix of (3.0.1) with $\lambda = \lambda_*$; and $\beta_{0,\boldsymbol{\Psi}}$ and $\beta_{1,\boldsymbol{\Psi}}$ are further abbreviated as β_0 and β_1, respectively.

Note that β_0 is the zero of $\psi_{12}\cos\beta - \psi_{22}\sin\beta$ on the interval $(0,\pi]$, and β_1 is that of $\psi_{11}\cos\beta - \psi_{21}\sin\beta$.

Lemma 3.1.2 Let $\lambda_* \in \mathbb{R}$, introduce $\boldsymbol{\Psi}$ and β_0 as in Definition 3.1.1, and define a function $u : (0,\pi] \to [0,\pi)$ by

$$u(\beta) = \operatorname{arccot}\frac{\psi_{11}\cos\beta - \psi_{21}\sin\beta}{-\psi_{12}\cos\beta + \psi_{22}\sin\beta}. \tag{3.1.13}$$

Then, the separated self-adjoint BCs having λ_* as an eigenvalue are

$$\mathbf{S}_{u(\beta),\beta}, \quad \beta \in (0,\pi]. \tag{3.1.14}$$

Moreover, there is an $n \in \mathbb{Z}$ such that n and $n+1$ are the only two choices for the index of λ_* as an eigenvalue for $\mathbf{S}_{u(\beta),\beta}$: the choice is n for $\beta_0 \leqslant \beta \leqslant \pi$, and $n+1$ for $0 < \beta < \beta_0$. It is understood that in (3.1.13), $u(\beta_0) = 0$.

Proof. The first claim is deduced from (4.8) in [49]. In the positive f case, the second claim comes from Theorem 2.2 in [104], and the proof there also works for the indefinite f case. □

We use the graph in Figure 3.1.1 to illustrate the results in the above lemma.

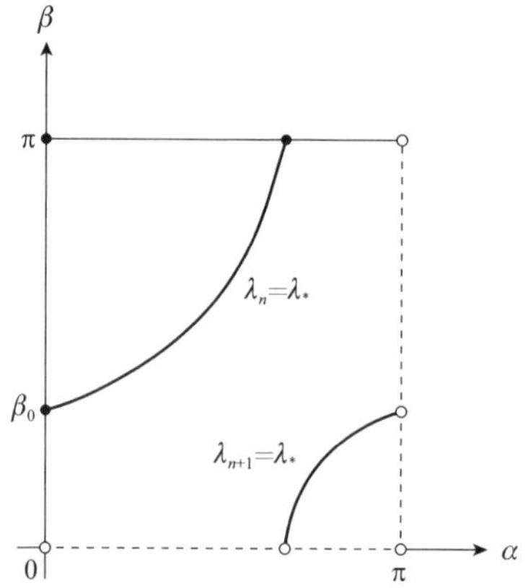

Figure 3.1.1 All separated self-adjoint BCs having λ_* as an eigenvalue

To give the indices of the eigenvalues for coupled self-adjoint BCs in the indefinite f case, we define

$$\mathcal{L}_1 = \{K \in \mathrm{SL}(2, \mathbb{R}); \ k_{11} > 0, \ k_{12} \leqslant 0\},$$
$$\mathcal{L}_2 = \{K \in \mathrm{SL}(2, \mathbb{R}); \ k_{11} \leqslant 0, \ k_{12} < 0\}.$$
(3.1.15)

Note that for each $K \in \mathrm{SL}(2, \mathbb{R})$, either $K \in \mathcal{L}_1 \cup \mathcal{L}_2$ or $-K \in \mathcal{L}_1 \cup \mathcal{L}_2$.

Lemma 3.1.3 Assume that f changes the sign in (a, b). Fix a $\gamma \in (-\pi, \pi)$ and a $K \in \mathrm{SL}(2, \mathbb{R})$, and set $\mathbf{A}$ to be the self-adjoint BC given by (3.1.7). Let $\{\mu_n; n \in \mathbb{Z}\}$ be the eigenvalues for $\mathbf{S}_{0,\beta_0,K}$, and denote by $\{\nu_n; n \in \mathbb{Z}\}$ the eigenvalues for $\mathbf{S}_{\pi/2,\beta_1,K}$.

(1) If $K \in \mathcal{L}_1 \cup \mathcal{L}_2$, then μ_n is not an eigenvalue for $\mathbf{A}$ for any odd $n \in \mathbb{Z}$; and for each odd $n \in \mathbb{Z}$, there are exactly two eigenvalues (counting multiplicity) for $\mathbf{A}$ in the interval (μ_n, μ_{n+2}), to be denoted by $\lambda_{n+1}(\mathbf{A})$ and $\lambda_{n+2}(\mathbf{A})$ in non-decreasing order.

(2) If $\gamma = 0$ and $K \in \mathrm{SL}(2, \mathbb{R}) \setminus (\mathcal{L}_1 \cup \mathcal{L}_2)$, then μ_n is not an eigenvalue for $\mathbf{A}$ for any even $n \in \mathbb{Z}$; and for each even $n \in \mathbb{Z}$, there are exactly two eigenvalues (counting multiplicity) for $\mathbf{A}$ in the interval (μ_n, μ_{n+2}), to be denoted by $\lambda_{n+1}(\mathbf{A})$ and $\lambda_{n+2}(\mathbf{A})$ in non-decreasing order.

In particular, there are infinitely many eigenvalues, unbounded from both below and above, for any coupled self-adjoint boundary condition.

Proof. See Theorem 2.17 in [98]. □

The following theorem gives a simple solution to the index problem for eigenvalues for coupled self-adjoint BCs, and it is true for both the positive f case and the indefinite

f case.

Theorem 3.1.1 Let λ_* be an eigenvalue for a coupled self-adjoint boundary condition $\mathbf{A}$, given by (3.1.7) with $\gamma \in \mathbb{R}$ and $\mathbf{K} \in \mathrm{SL}(2,\mathbb{R})$. Introduce $\beta_{0,\mathbf{K}}, \beta_{1,\mathbf{K}}, \Psi, \beta_0$ and β_1 by Definition 3.1.1.

(1) We have that $\beta_{0,\mathbf{K}} \neq \beta_{1,\mathbf{K}}$, and $\beta_0 \neq \beta_1$.

(2) The number λ_* is also an eigenvalue for the separated self-adjoint BC $\mathbf{S}_{0,\beta_0}$. Denote by $n \in \mathbb{Z}$ the corresponding index.

(3) If $\beta_{0,\mathbf{K}} > \beta_0$, then λ_* is a simple eigenvalue for $\mathbf{A}$, and its index is n.

(4) If $\beta_{0,\mathbf{K}} < \beta_0$, then λ_* is a simple eigenvalue for $\mathbf{A}$, and its index is $n+1$.

(5) If $\beta_{0,\mathbf{K}} = \beta_0$ and $\beta_{1,\mathbf{K}} = \beta_1$, then λ_* is a double eigenvalue for $\mathbf{A}$, and its indices are n and $n+1$.

(6) If $\beta_{0,\mathbf{K}} = \beta_0$ and either

$$\beta_{1,\mathbf{K}} > \beta_1 > \beta_0 \quad \text{or} \quad \beta_1 > \beta_0 > \beta_{1,\mathbf{K}} \quad \text{or} \quad \beta_0 > \beta_{1,\mathbf{K}} > \beta_1, \tag{3.1.16}$$

then λ_* is a simple eigenvalue for $\mathbf{A}$, and its index is n.

(7) If $\beta_{0,\mathbf{K}} = \beta_0$ and either

$$\beta_1 > \beta_{1,\mathbf{K}} > \beta_0 \quad \text{or} \quad \beta_{1,\mathbf{K}} > \beta_0 > \beta_1 \quad \text{or} \quad \beta_0 > \beta_1 > \beta_{1,\mathbf{K}}, \tag{3.1.17}$$

then λ_* is a simple eigenvalue for $\mathbf{A}$, and its index is $n+1$.

Note that by (1), all possibilities for the relations among $\beta_{0,\mathbf{K}}, \beta_{1,\mathbf{K}}, \beta_0$ and β_1 are covered in (3)–(7).

Proof. See Theorem 1.16 in [100]. □

Using Lemma 3.1.2, the solution to the index problem given in Theorem 3.1.1 can be restated as follows: if either $\beta_{0,\mathbf{K}} > \beta_0$, or $\beta_{0,\mathbf{K}} = \beta_0$ and one set of the conditions in (3.1.16) are satisfied, then λ_* is a simple eigenvalue for $\mathbf{A}$, and its index is equal to the index of λ_* as an eigenvalue for $\mathbf{S}_{u(\beta),\beta}$ with any $\beta \in [\beta_0, \pi]$; if either $\beta_{0,\mathbf{K}} < \beta_0$, or $\beta_{0,\mathbf{K}} = \beta_0$ and one set of the requirements in (3.1.17) are fulfilled, then λ_* is a simple eigenvalue for $\mathbf{A}$, and its index is equal to the index of λ_* as an eigenvalue for $\mathbf{S}_{u(\beta),\beta}$ with any $\beta \in (0, \beta_0)$; if $\beta_{0,\mathbf{K}} = \beta_0$ and $\beta_{1,\mathbf{K}} = \beta_1$, then λ_* is a double eigenvalue for $\mathbf{A}$, and its smaller index is equal to the index of λ_* as an eigenvalue for $\mathbf{S}_{u(\beta),\beta}$ with any $\beta \in [\beta_0, \pi]$. Therefore, the determination of the index or indices of an eigenvalue for a coupled self-adjoint BC is converted to finding the index of the same number as an eigenvalue for one separated self-adjoint BC.

A direct consequence of Theorem 3.1.1 is as follows.

Corollary 3.1.1 Let λ_*, **A**, Ψ, β_0 and n be the same as in Theorem 3.1.1. If λ_* is double for **A**, then its indices are n and $n+1$.

Remark 3.1.1 The following is a general procedure for applying Theorem 3.1.1 to concrete index problems: if λ_* is an eigenvalue for the coupled self-adjoint BC **A** given by (3.1.7) with an eigenfunction y_*, then

First, using the reduction of order method, one can find another solution z_* to (3.0.1) with $\lambda = \lambda_*$ linearly independent of y_* (for example, in each subinterval where f is constant,

$$z_* = y_* \int \frac{1}{y_*^2} \qquad (3.1.18)$$

is such a solution). Second, from y_* and z_* we can obtain $\Phi(x, \lambda_*)$ and hence Ψ, β_0 and β_1. Third, since $\phi_{12}(x, \lambda_*)$ is an eigenfunction for λ_* as an eigenvalue for the separated self-adjoint BC $\mathbf{S}_{0,\beta_0}$, the corresponding index n can be obtained from $\phi_{12}(x, \lambda_*)$ via (3.1.9) (or simply a zero count when f is positive). Fourth, compare $\beta_{0,K}, \beta_{1,K}, \beta_0$ and β_1 and apply Theorem 3.1.1 to obtain the index or indices of λ_* as an eigenvalue for **A**.

Note that y_* has only isolated zeros on $[a, b]$; and by the uniqueness of solutions to linear ordinary differential equations, z_* given by (3.1.18) and fz'_* have continuous extensions to these points. However, when an endpoint of (a, b) is a zero of y_*, some work (such as integration by parts) has to be done to evaluate z_* and fz'_* at that endpoint to determine $\Phi(x, \lambda_*)$ and Ψ.

Theoretically, Theorem 3.1.1 gives a complete solution to the index problem; however, the following results, slightly more general than Theorem 3.1.1 (6) and (7), are useful in practical computations. Here, for any two objects c_1 and c_2, the notation $\{c_1, c_2\}$ with bold-faced braces means each of c_1 and c_2.

Corollary 3.1.2 Let notation be the same as in Theorem 3.1.1.
(1) If one of

$$\beta_{1,K} > \beta_1 > \{\beta_0, \beta_{0,K}\}, \quad \beta_1 > \{\beta_0, \beta_{0,K}\} > \beta_{1,K}, \quad \{\beta_0, \beta_{0,K}\} > \beta_{1,K} > \beta_1$$
(3.1.19)

is true, then λ_* is a simple eigenvalue for **A**, and its index is n.
(2) If one of

$$\beta_1 > \beta_{1,K} > \{\beta_0, \beta_{0,K}\}, \quad \beta_{1,K} > \{\beta_0, \beta_{0,K}\} > \beta_1, \quad \{\beta_0, \beta_{0,K}\} > \beta_1 > \beta_{1,K}$$
(3.1.20)

is true, then λ_* is a simple eigenvalue for **A**, and its index is $n+1$.

Proof. See Corollary 2.25 in [100]. □

It is natural to ask: when do we have $\beta_{0,K} = \beta_0$? The following result gives a complete description of this situation, and it also holds for both the positive f case and the indefinite f case.

Theorem 3.1.2 Let $\lambda_*, \gamma, K, A, \beta_{0,K}, \Psi$ and β_0 be the same as above. Assume further that γ is in its standard range $[0, \pi)$. Then, $\beta_{0,K} = \beta_0$ if and only if $\gamma = 0$ and the second columns of K and Ψ are equal, i.e.,

$$e^{i\gamma} K = K = \begin{pmatrix} k_{11} & \psi_{12} \\ k_{21} & \psi_{22} \end{pmatrix}. \tag{3.1.21}$$

Proof. See Theorem 1.20 in [100]. □

The sufficiency in Theorem 3.1.2 can be seen easily: when $e^{i\gamma} K$ is given by (3.1.21), $\beta_{0,K} = \beta_0$ by definition, and λ_* is an eigenvalue for the coupled BC **A** since $\phi_{12}(\cdot, \lambda_*)$ is an eigenfunction; however, proof of the necessity requires some work.

3.2 Algorithm and implementation

Given an eigenvalue λ_* for a coupled self-adjoint BC **A**, Theorem 3.1.1 yields the following algorithm for computing the index or indices of λ_*.

Step 1. Normalize **A** into the form given in (3.1.7) with $\gamma \in \mathbb{R}$ and $K \in \mathrm{SL}(2, \mathbb{R})$; and compute $\beta_{0,K}, \beta_{1,K} \in (0, \pi]$ by (3.1.11).

Step 2. Approximate the fundamental solution matrix $\Phi(x, \lambda_*)$ for $a \leqslant x \leqslant b$ using

$$\Phi'(x, \lambda_*) = \begin{pmatrix} 0 & 1/f(x) \\ q(x) - \lambda_* w(x) & 0 \end{pmatrix} \Phi(x, \lambda_*), \quad \Phi(a, \lambda_*) = I; \tag{3.2.1}$$

define the transfer matrix Ψ via (3.1.12); and compute $\beta_0, \beta_1 \in (0, \pi]$ by

$$\begin{aligned} \tan \beta_0 &= \psi_{12}/\psi_{22} \quad \text{or} \quad \cot \beta_0 = \psi_{22}/\psi_{12}, \\ \tan \beta_1 &= \psi_{11}/\psi_{21} \quad \text{or} \quad \cot \beta_1 = \psi_{21}/\psi_{11}. \end{aligned} \tag{3.2.2}$$

Step 3. Let

$$\beta_* = \begin{cases} (\beta_0 + \pi)/2 & \text{if } \beta_0 \leqslant \pi/2, \\ \beta_0/2 & \text{otherwise;} \end{cases} \tag{3.2.3}$$

and compute $\alpha_* \in [0, \pi)$ by

$$\tan\alpha_* = \frac{-\psi_{12}\cos\beta_* + \psi_{22}\sin\beta_*}{\psi_{11}\cos\beta_* - \psi_{21}\sin\beta_*} \quad \text{or} \quad \cot\alpha_* = \frac{\psi_{11}\cos\beta_* - \psi_{21}\sin\beta_*}{-\psi_{12}\cos\beta_* + \psi_{22}\sin\beta_*}.$$
(3.2.4)

Step 4. Use $\Phi(x, \lambda_*)$ to form an eigenfunction y_* for λ_* as an eigenvalue for the separated BC S_{α_*,β_*}; and compute the "change" $\theta_*(b, \lambda_*) - \beta_*$ of the Prüfer angle θ_* of y_* on the interval $[a, b]$ (here quotation marks are used since the initial value of θ_* is α_*, not β_*). Denote the change by $(n-1)\pi$ if $\beta_0 \leqslant \pi/2$, and by $n\pi$ otherwise.

Step 5. Use Theorem 3.1.1 to determine the index or indices of λ_* as an eigenvalue for $\mathbf{A}$: the index is n if $\beta_{0,K} > \beta_0$; the index is $n+1$ if $\beta_{0,K} < \beta_0$; the index is n if $\beta_{0,K} = \beta_0$ and one set of the conditions in (3.1.16) are satisfied; the index is $n+1$ if $\beta_{0,K} = \beta_0$ and one set of the conditions in (3.1.17) are fulfilled; and the indices are n and $n+1$ if $\beta_{0,K} = \beta_0$ and $\beta_{1,K} = \beta_1$.

Next, we describe in some detail an implementation in Mathematica of the above algorithm. In this implementation, the case of separated self-adjoint BCs is handled at the same time, and we assume that a sequence of consecutive eigenvalues (not necessarily a single eigenvalue only) together with their multiplicities is given.

3.2.1 Piecewise constant case

```
Input: npieces, a[i], f[i], q[i], w[i], mAB, nterms, ev[i], mt[i]
Optional input: psiYY, inpYY, czYY
Output: print outs,beta0YY,beta1YY,beta0KYY,beta1KYY,indexYY[i]
```

Here, the number npieces of pieces and the sequences a[i], f[i], q[i] and w[i] of real numbers specify the piecewise constant S-L equation, with i running from 0 to npieces in a[i] and from 1 to npieces in f[i], q[i] and w[i]; mAB is the 2×4 coefficient matrix of the BC; and the given partial sequence of consecutive eigenvalues ev[i] has multiplicities mt[i], where i varies from 1 to nterms.

First, the self-adjointness of the given BC $[A \mid B]$ is checked: if all the 2×2 matrices consisting of the first and third columns, of the first and fourth columns, of the second and third columns, and of the second and fourth columns of $(A \mid B)$ are singular, then the message

"The boundary condition is really NOT self-adjoint!"

is printed out, and the execution of the codes ends; and if the self-adjointness condition

$$A \begin{pmatrix} 0 & -1 \\ 1 & 0 \end{pmatrix} A^* = B \begin{pmatrix} 0 & -1 \\ 1 & 0 \end{pmatrix} B^* \quad (3.2.5)$$

is not satisfied, then the message

"The boundary condition is NOT self-adjoint!"

comes out, and the execution stops.

Second, the BC is normalized: it is written in its real form if it is separated, and into the form in (3.1.7) with $\gamma \in (-\pi, \pi]$ and $\boldsymbol{K} \in \mathrm{SL}(2, \mathbb{R})$ if the BC is coupled; and $\beta_{0,K}$ and $\beta_{1,K}$ are computed via (3.1.11) in the latter case. Here the interval $(-\pi, \pi]$ for the values of γ is caused by the Arg function in Mathematica.

Third, if the optional $\mathbb{N}$-valued parameter psiYY has a given value, then enter ev[psiYY] into λ_*; and if psiYY is not given, set λ_* to the eigenvalue in the partial sequence ev[i] with the smallest absolute value.

Fourth, if the BC is separated, then its coefficients at a are written as $\cos \alpha$ and $-\sin \alpha$; and if the BC is coupled, then β_0, β_1, β_* and α_* are computed via (3.2.2)–(3.2.4). In the latter case, when $\beta_0 \in (0, 0.0001) \cup (0.9999\pi, \pi]$, a warning message is given:

"Warning: since $\beta_0 = \cdots$ is close to 0 or π,"

"use a different eigenvalue to verify the indices."

We set $\beta_0 = \beta_{0,K}$ and $\beta_1 = \beta_{1,K}$ when the multiplicity mt[psiYY] of λ_* is 2, to avoid possible mistakes caused by an inaccurate β_0 close to 0 or π coming from the error (even a very small one) in λ_* (remember that λ_* is not exact, in general). To use a specific eigenvalue (say, the third term) in the partial sequence ev[i], we simply enter the corresponding natural number (i.e., 3 in our example) into psiYY.

Fifth, an eigenfunction for λ_*, as an eigenvalue for the BC if it is a separated one and $\mathbf{S}_{\alpha_*,\beta_*}$ if the BC is a coupled one, is formed using the file Phi-matrix-pc; and the change in the Prüfer angle of the eigenfunction is computed. For the latter task, we find the Prüfer angle at a few points on $[a, b]$ and compute the change, then divide each subinterval into two and compute the change again, and repeat the dividing procedure until we get the same change for three consecutive times. The initial number of points used can be controlled via the optional $\mathbb{Z}$-valued parameter inpYY, whose default value and minimum value are both 9. When f is positive, the change can be verified using the number of zeros of the eigenfunction in the open interval (a, b). To activate this verification, we give a positive value to the optional $\mathbb{R}$-valued parameter czYY, whose default value is -1.

Finally, the index of λ_* as an eigenvalue for the separated BC is determined from the change in the Prüfer angle; the index or indices of λ_* as an eigenvalue for the given BC is computed according to Theorem 3.1.1 if the BC is a coupled one; and the indices of all eigenvalues in ev[i] are computed and printed out, and the index or smaller index of each ev[i] is saved in indexYY[i]. For a coupled BC, if λ_* is not double and

$$|\beta_{0,K} - \beta_0| < 10^{-3}, \qquad \beta_{1,K} \notin [c,d], \qquad \beta_1 \notin [c,d], \qquad (3.2.6)$$

where $c = \min\{\beta_{0,K}, \beta_0\}$ and $d = \max\{\beta_{0,K}, \beta_0\}$, then Corollary 3.1.2 is used to avoid possible mistakes caused by the error (again, even a very small one) in the approximation of β_0.

3.2.2 General case

```
input: aa, bb, f[t_], q[t_], w[t_], mAB, nterms, ev[i], mt[i]
Optional input: psiYY, inpYY, czYY
Output: print outs,beta0YY,beta1YY,beta0KYY,beta1KYY,indexYY[i]
```

Here, the numbers aa and bb are the approximations of the endpoints a and b, respectively, used in actual calculations (to handle the case where one of $1/f$, q and w is unbounded at a or b). The meaning of the other symbols is either obvious or as before.

The implementation in this general case is almost the same as the piecewise constant case except that to get an approximation of $\Phi(x, \lambda_*)$, the file Phi-matrix-rglr is called, instead of Phi-matrix-pc.

Remark 3.2.1 From the discussions of this section we see that when the exact value of β_0 is not in the extreme case $\beta_0 = \pi$, the index or indices of a given approximate eigenvalue λ_* can be correctly computed from an approximation of $\Phi(b, \lambda_*)$, as long as the approximations of λ_* and $\Phi(b, \lambda_*)$ are sufficiently accurate. For the extreme case $\beta_0 = \pi$, the codes only suggest the use of another eigenvalue. Generically, this is certainly enough already. Moreover, a complete solution of the extreme case will be given in a forthcoming publication.

3.3 Examples with a positive f

In this section, we give explicit S-L problems to show that in the positive f case, each of the 9 possibilities in Theorem 3.1.1 can happen. We also present a couple of examples with a positive f where the first few eigenvalues are not in the chosen computation range. In all these examples, we will always use S-L problems with an exact eigenvalue 1. This choice of the exact eigenvalue is not special: it can be shifted to any real number by adding a corresponding multiple of w to q.

Example 3.3.1 Let $q(x) = x - 1$, and $w(x) = x$. Then, 1 is an eigenvalue of the

S-L problem

$$-y'' + qy = \lambda wy \text{ in } (0, \pi), \qquad Y(\pi) = \begin{pmatrix} -1 & -1 \\ 0 & -1 \end{pmatrix} Y(0) \qquad (3.3.1)$$

with an eigenfunction $y(x) = \cos x$. It is direct to verify that $\sin x$ is another solution to the S-L equation in (3.3.1) with $\lambda = 1$, and hence

$$\Phi(x, 1) = \begin{pmatrix} \cos x & \sin x \\ -\sin x & \cos x \end{pmatrix}, \qquad \Psi = \Phi(\pi, 1) = \begin{pmatrix} -1 & 0 \\ 0 & -1 \end{pmatrix}. \qquad (3.3.2)$$

Note that $\sin x$ has no zeros in $(0, \pi)$. Thus, by Remark 3.1.1 and (3.3.2), the index of 1 as an eigenvalue for the BC $\mathbf{S}_{0,\beta_0}$ is 1. Since

$$\beta_0 = \pi > \beta_{0,K} = \pi/4, \qquad (3.3.3)$$

$\lambda_2 = 1$ by Theorem 3.1.1. In fact, the codes obtain the following approximations of the eigenvalues on the interval $[-10, 10]$:

$$-7.93477, \quad 1.0, \quad 1.57704, \quad 6.69360, \quad 9.47920, \qquad (3.3.4)$$

all eigenvalues are simple; while the psiYY= 2 then give their indices:

$$1, \quad 2, \quad 3, \quad 4, \quad 5, \qquad (3.3.5)$$

and the message "Warning: since β_0 = 3.14159 is close to 0 or π, use a different eigenvalue to verify the indices". Moreover, if one of 1, 3, 4 and 5 is entered into psiYY, then one obtains the same index sequence as (3.3.5).

Example 3.3.2 Let $q(x) = x - 1$, and $w(x) = x$. Then, 1 is an eigenvalue of the S-L problem

$$-y'' + qy = \lambda wy \text{ in } (0, 7\pi/4), \qquad Y(7\pi/4) = \frac{1}{\sqrt{2}} \begin{pmatrix} 1 & -2 \\ 1 & 0 \end{pmatrix} Y(0) \qquad (3.3.6)$$

with an eigenfunction $y(x) = \cos x$. It is direct to verify that $\sin x$ is another solution to the S-L equation in (3.3.6) with $\lambda = 1$, and hence

$$\Phi(x, 1) = \begin{pmatrix} \cos x & \sin x \\ -\sin x & \cos x \end{pmatrix}, \qquad \Psi = \Phi(7\pi/4, 1) = \frac{1}{\sqrt{2}} \begin{pmatrix} 1 & -1 \\ 1 & 1 \end{pmatrix}. \qquad (3.3.7)$$

Note that $\sin x$ has exactly one zero in $(0, 7\pi/4)$. Thus, the index of 1 as an eigenvalue for $\mathbf{S}_{0,\beta_0}$ is 2. Since

$$\beta_0 = 3\pi/4 > \beta_{0,K} = \pi/2, \qquad (3.3.8)$$

$\lambda_3=1$ by Theorem 3.1.1. Actually, the codes with psiYY=3 obtain the following approximations of the eigenvalues on the interval $[-4, 4]$ and their indices

$$\lambda_1 \approx 0.24995, \quad \lambda_2 \approx 0.90578, \quad \lambda_3 \approx 0.99999,$$
$$\lambda_4 \approx 2.04145, \quad \lambda_5 \approx 2.61244. \tag{3.3.9}$$

Therefore, the numerical results are consistent with Theorem 3.1.1.

If the BC chosen is

$$\mathbf{Y}(7\pi/4) = \begin{pmatrix} 1/\sqrt{2} & -\sqrt{2}/(1+\sqrt{3}) \\ 1/\sqrt{2} & \sqrt{6}/(1+\sqrt{3}) \end{pmatrix} \mathbf{Y}(0), \tag{3.3.10}$$

then

$$\beta_{0,K} = 5\pi/6 > \beta_0 = 3\pi/4, \tag{3.3.11}$$

thus $\lambda_2 = 1$ by Theorem 3.1.1. In fact, the codes with psiYY=2 yield the following approximations of the eigenvalues on the interval $[-5, 5]$ and their indices

$$\lambda_1 \approx 0.28841, \quad \lambda_2 \approx 1.0, \quad \lambda_3 \approx 1.03071,$$
$$\lambda_4 \approx 2.30948, \quad \lambda_5 \approx 2.61988, \quad \lambda_6 \approx 4.63613. \tag{3.3.12}$$

If we change the BC to

$$\mathbf{Y}(7\pi/4) = \frac{1}{\sqrt{2}} \begin{pmatrix} 1 & -1 \\ 1 & 1 \end{pmatrix} \mathbf{Y}(0), \tag{3.3.13}$$

then

$$\beta_0 = \beta_{0,K} = 3\pi/4, \quad \beta_1 = \beta_{1,K} = \pi/4. \tag{3.3.14}$$

Thus, $\lambda_2=\lambda_3=1$ by Theorem 3.1.1. Actually, the codes with psiYY=2 obtain the following approximations of the eigenvalues on the interval $[-5, 5]$ and their indices

$$\lambda_1 \approx 0.27973, \quad \lambda_2 = \lambda_3 \approx 0.99985, \quad \lambda_4 \approx 2.23403,$$
$$\lambda_5 \approx 2.61682, \quad \lambda_6 \approx 4.52107. \tag{3.3.15}$$

Let $b \in (0, \pi)$. Then, for $\lambda \neq 0$, the fundamental solution matrix to the Fourier equation $-y'' = \lambda y$ in $(0, b)$ is

$$\Phi(x, \lambda) = \begin{pmatrix} \cos(\sqrt{\lambda}\, x) & \frac{1}{\sqrt{\lambda}} \sin(\sqrt{\lambda}\, x) \\ -\sqrt{\lambda} \sin(\sqrt{\lambda}\, x) & \cos(\sqrt{\lambda}\, x) \end{pmatrix}. \tag{3.3.16}$$

Thus, for the special value $\lambda_* = 1$ of λ,

$$\boldsymbol{\Psi} = \boldsymbol{\Phi}(b, 1) = \begin{pmatrix} \cos b & \sin b \\ -\sin b & \cos b \end{pmatrix}, \qquad (3.3.17)$$

and hence $\beta_0 = b$, while $\beta_1 = b + \pi/2$ if $b \in (0, \pi/2]$ and $b - \pi/2$ if $b \in (\pi/2, \pi)$. The next six examples (Example 3.3.3-3.3.8) use these observations. Note that $\sin x$ has no zero in $(0, b)$. Thus, the index of 1 as an eigenvalue for $\mathbf{S}_{0,\beta_0}$ is always 1.

Example 3.3.3 Consider the S-L problem

$$-y'' = \lambda y \text{ in } (0, \pi/4), \qquad \boldsymbol{Y}(\pi/4) = \frac{1}{\sqrt{2}} \begin{pmatrix} 0 & 1 \\ -2 & 1 \end{pmatrix} \boldsymbol{Y}(0) \qquad (3.3.18)$$

associated with the Fourier equation. Then, 1 is an eigenvalue by Theorem 3.1.2 and (3.3.17). In this example, $\lambda_* = 1.0$ is used by the codes, and hence

$$\beta_{1,K} = \pi > \beta_1 = 3\pi/4 > \beta_{0,K} = \beta_0 = \pi/4. \qquad (3.3.19)$$

Thus, $\lambda_1 = 1$ by Theorem 3.1.1. Actually, the codes yield the following approximations of the eigenvalues on the interval $[-200, 200]$ and their indices

$$\lambda_1 \approx 1.0, \quad \lambda_2 \approx 5.41211, \quad \lambda_3 \approx 68.5423, \quad \lambda_4 \approx 134.183. \qquad (3.3.20)$$

For the S-L problem (3.3.18), if we use a λ_* sufficiently close to 1, but <1, then β_0 is close to $\pi/4 = \beta_{0,K}$, but $> \pi/4$. For example, if $\lambda_* = 0.999999$ is used, then $\beta_0 \approx 0.7854$. In such a situation, if Corollary 3.1.2 were not used in the codes, a wrong index (i.e., 2) would be found for λ_*. This demonstrates the importance of Corollary 3.1.2.

Example 3.3.4 Take the S-L problem

$$-y'' = \lambda y \text{ in } (0, \pi/4), \qquad \boldsymbol{Y}(\pi/4) = \begin{pmatrix} \sqrt{2}/(1-\sqrt{3}) & 1/\sqrt{2} \\ \sqrt{6}/(1-\sqrt{3}) & 1/\sqrt{2} \end{pmatrix} \boldsymbol{Y}(0) \qquad (3.3.21)$$

associated with the Fourier equation. Then, 1 is an eigenvalue. In this example, $\lambda_* = 1.0$ is used by codes, and hence

$$\beta_1 = 3\pi/4 > \beta_{0,K} = \beta_0 = \pi/4 > \beta_{1,K} = \pi/6. \qquad (3.3.22)$$

Thus, $\lambda_1 = 1$ by Theorem 3.1.1. In fact, the codes find the following approximations of the eigenvalues on the interval $[-200, 200]$ and their indices

$$\lambda_1 \approx 1.0, \quad \lambda_2 \approx 13.7301, \quad \lambda_3 \approx 74.4638, \quad \lambda_4 \approx 141.265. \qquad (3.3.23)$$

Example 3.3.5 Pick the S-L problem

$$-y'' = \lambda y \text{ in } (0, 3\pi/4), \qquad Y(3\pi/4) = \frac{1}{\sqrt{2}} \begin{pmatrix} -2 & 1 \\ 0 & -1 \end{pmatrix} Y(0) \qquad (3.3.24)$$

associated with the Fourier equation. Then, 1 is an eigenvalue. In this example, $\lambda_* = 1.0$ is used by the codes, and hence

$$\beta_{0,K} = \beta_0 = 3\pi/4 > \beta_{1,K} = \pi/2 > \beta_1 = \pi/4. \qquad (3.3.25)$$

Thus, $\lambda_1 = 1$ by Theorem 3.1.1. Actually, the codes obtain the following approximations of the eigenvalues on the interval $[-40, 40]$ and their indices

$$\lambda_1 \approx 1.0, \quad \lambda_2 \approx 1.90575, \quad \lambda_3 \approx 10.7223, \quad \lambda_4 \approx 16.1431, \quad \lambda_5 \approx 32.8343.$$
$$(3.3.26)$$

Example 3.3.6 Consider the S-L problem

$$-y'' = \lambda y \text{ in } (0, \pi/4), \qquad Y(\pi/4) = \frac{1}{\sqrt{2}} \begin{pmatrix} 2 & 1 \\ 0 & 1 \end{pmatrix} Y(0) \qquad (3.3.27)$$

associated with the Fourier equation. Then, 1 is an eigenvalue. In this example, $\lambda_* = 1.0$ is used by the codes, and hence

$$\beta_1 = 3\pi/4 > \beta_{1,K} = \pi/2 > \beta_{0,K} = \beta_0 = \pi/4. \qquad (3.3.28)$$

Thus, $\lambda_2 = 1$ by Theorem 3.1.1. In fact, the codes yield the following approximations of the eigenvalues on the interval $[-200, 200]$ and their indices

$$\lambda_1 \approx -4.42279, \quad \lambda_2 \approx 1.0, \quad \lambda_3 \approx 63.5641, \quad \lambda_4 \approx 129.091. \qquad (3.3.29)$$

Example 3.3.7 Take the S-L problem

$$-y'' = \lambda y \text{ in } (0, 3\pi/4), \qquad Y(3\pi/4) = \frac{1}{\sqrt{2}} \begin{pmatrix} 0 & 1 \\ -2 & -1 \end{pmatrix} Y(0) \qquad (3.3.30)$$

associated with the Fourier equation. Then, 1 is an eigenvalue. In this example, $\lambda_* = 1.0$ is used by the codes, and hence

$$\beta_{1,K} = \pi > \beta_{0,K} = \beta_0 = 3\pi/4 > \beta_1 = \pi/4. \qquad (3.3.31)$$

Thus, $\lambda_2 = 1$ by Theorem 3.1.1. Actually, the codes with psiYY= 2 find the following approximations of the eigenvalues on the interval $[-40, 40]$ and their indices

$$\lambda_1 \approx -0.45433, \quad \lambda_2 \approx 1.0, \quad \lambda_3 \approx 9.74718, \quad \lambda_4 \approx 14.4665, \quad \lambda_5 \approx 31.4752.$$
$$(3.3.32)$$

Example 3.3.8 Pick the S-L problem

$$-y'' = \lambda y \text{ in } (0, 3\pi/4), \, Y(3\pi/4) = \begin{pmatrix} -\sqrt{2}/(1+\sqrt{3}) & 1/\sqrt{2} \\ -\sqrt{6}/(1+\sqrt{3}) & -1/\sqrt{2} \end{pmatrix} Y(0) \quad (3.3.33)$$

associated with the Fourier equation. Then, 1 is an eigenvalue. In this example, $\lambda_* = 1.0$ is used by the codes, and hence

$$\beta_{0,K} = \beta_0 = 3\pi/4 > \beta_1 = \pi/4 > \beta_{1,K} = \pi/6. \quad (3.3.34)$$

Thus, $\lambda_2 = 1$ by Theorem 3.1.1. In fact, the codes with psiYY= 2 obtain the following approximations of the eigenvalues on the interval $[-40, 40]$ and their indices

$$\lambda_1 \approx 0.67676, \quad \lambda_2 \approx 1.0, \quad \lambda_3 \approx 10.1911, \quad \lambda_4 \approx 15.0625, \quad \lambda_5 \approx 32.0167.$$
$$(3.3.35)$$

Example 3.3.9 Let $q(x) = x - 16$, $w(x) = x$, and $\delta \in (0, \pi/2]$. Consider the S-L problem

$$-y'' + qy = \lambda w y \text{ in } (0, \pi/2 + \delta/4), \quad AY(0) + BY(\pi/2 + \delta/4) = 0, \quad (3.3.36)$$

where

$$[A \,|\, B] = \begin{bmatrix} 1 & t & 0 & -it\cos\delta \\ 0 & it\cos\delta & -1 & t \end{bmatrix} \quad (3.3.37)$$

with

$$t = \frac{\sin\delta}{4(\sqrt{\sin^4\delta + \cos^2\delta} + \cos\delta)}. \quad (3.3.38)$$

Then, $\sin 4x$ and $\cos 4x$ are two solutions to the S-L equation in (3.3.36) with $\lambda = 1$, and hence

$$\Phi(x, 1) = \begin{pmatrix} \cos 4x & (\sin 4x)/4 \\ -4\sin 4x & \cos 4x \end{pmatrix},$$

$$\Psi = \Phi(\pi/2 + \delta/4, 1) - \begin{pmatrix} \cos\delta & \sin(\delta/4) \\ -4\sin\delta & \cos\delta \end{pmatrix}. \quad (3.3.39)$$

Direct calculations using the characteristic function Δ and the matrix Ψ yield that 1 is an eigenvalue. Note that $\sin 4x$ has exactly two zeros on $(0, \pi/2 + \delta/4)$. Thus, the index of 1 as an eigenvalue for $\mathbf{S}_{0,\beta_0}$ is 3.

Since $[A \,|\, B] = [i\mathbf{K} \,|\, -I]$, where

$$K = -\frac{1}{t\cos\delta}\begin{pmatrix} t & t^2\sin^2\delta \\ 1 & t \end{pmatrix}, \quad (3.3.40)$$

we have that

$$\beta_0 = \operatorname{arccot}\frac{4\cos\delta}{\sin\delta} > \beta_{0,K} = \operatorname{arccot}\frac{4(\sqrt{\sin^4\delta + \cos^2\delta} + \cos\delta)}{\sin^3\delta}, \quad (3.3.41)$$

and hence $\lambda_4 = 1$ by Theorem 3.1.1. We remark that this is true for all $\delta \in (0, \pi/2]$.

If $\delta = \pi/3$, then the codes yield the following approximations of the eigenvalues on the interval $[-500, 50]$ and their indices

$$\lambda_2 \approx -35.0185, \quad \lambda_3 \approx -12.9914, \quad \lambda_4 \approx 0.99999, \quad \lambda_5 \approx 21.9230. \quad (3.3.42)$$

Using the codes, one can also see that $\lambda_1 \approx -2296.98$.

If $\delta = \pi/20$, then the codes obtain the following approximations of the eigenvalues on the interval $[-1000, 50]$ and their indices

$$\lambda_3 \approx -16.1334, \quad \lambda_4 \approx 0.99999, \quad \lambda_5 \approx 26.7713. \quad (3.3.43)$$

Using the codes, one can also see that $\lambda_1 \approx -197{,}966$, $\lambda_2 \approx -1{,}205.23$.

This example demonstrates an important feature of Theorem 3.1.1, i.e., the index or indices of a given eigenvalue can be determined without any information about the eigenvalues below the given eigenvalue. It is essential when (approximations of) the first few eigenvalues are not available. Actually, when $\lambda < -100$, the solutions to the S-L equation in (3.3.36) given by the ODE solver in Mathematica can not reach any reasonable accuracy. Therefore, it is difficult to find good approximations of λ_1 for $\delta \leq \pi/3$, and even of λ_2 for $\delta \leq \pi/10$. Figure 3.3.1 below shows that for $i = 1$ and 2, $\lambda_i \to -\infty$ as $\delta \to 0^+$.

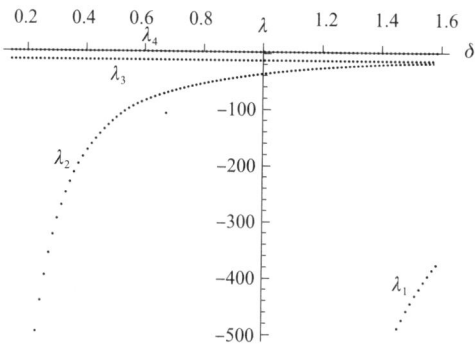

Figure 3.3.1 The eigenvalues in the interval $[-500, 2]$ for $\pi/20 \leq \delta \leq \pi/2$

Example 3.3.10 Let $k \in \mathbb{C}$, and set $q(x) = k^2\sin^2 x - 3k\cos x$. Then, direct calculations show that $\lambda_* = 1$ is an eigenvalue of the S-L problem

$$-y'' + qy = \lambda y \text{ in } (0, 2\pi), \qquad \mathbf{Y}(2\pi) = \mathbf{Y}(0) \tag{3.3.44}$$

with an eigenfunction $y_*(x) = e^{k\cos x} \sin x$; while codes find the following approximations of the eigenvalues on the interval $[-2, 2]$ and their indices

$$\begin{aligned}
k &= \pm 1: & \lambda_1 &\approx -1.56155, & \lambda_2 &\approx 0.999983; \\
k &= \pm 0.75: & \lambda_1 &\approx -1.08114, & \lambda_2 &\approx 0.999983; \\
k &= \pm 0.5: & \lambda_1 &\approx -0.618063, & \lambda_2 &\approx 0.999983, & \lambda_3 &\approx 1.61805; \\
k &= \pm 0.25: & \lambda_1 &\approx -0.207082, & \lambda_2 &\approx 0.999983, & \lambda_3 &\approx 1.20713; \\
k &= 0: & \lambda_1 &\approx -0.0, & \lambda_2 &= \lambda_3 \approx 0.999983.
\end{aligned} \tag{3.3.45}$$

This example has been discussed by Fulton in [99], where he conjectured that for $k \in \mathbb{R} \setminus \{0\}$, $\lambda_* = 1$ is simple and its index is 2. Therefore, our numerical examples support his conjecture.

3.4 Examples with an indefinite f

We now give 9 examples to demonstrate that in the case where f changes the sign, all 9 possible relations among $\beta_0, \beta_1, \beta_{0,K}$ and $\beta_{1,K}$ in Theorem 3.1.1 are also realized in some explicit situations. Again, the codes are used to find approximate eigenvalues; and all S-L problems are treated as general regular problems. Moreover, we will always use S-L problems with an exact eigenvalue 1.

In this section, we always set $f(x) = \text{sgn} x$.

Example 3.4.1 Let $q(x) = x^2 - 9\text{sgn}x$, and $w(x) = x^2$. Then, 1 is an eigenvalue of the S-L problem

$$-(fy')' + qy = \lambda w y \text{ in } (-3\pi/2, \pi/4), \qquad \mathbf{Y}(\pi/4) = \mathbf{K}\mathbf{Y}(-3\pi/2) \tag{3.4.1}$$

with an eigenfunction $y(x) = \text{sgn}x \sin 3x$, where

$$\mathbf{K} = \sqrt{2} \begin{pmatrix} 1/2 & -1/[3(\sqrt{3}-1)] \\ -3/2 & \sqrt{3}/(\sqrt{3}-1) \end{pmatrix}. \tag{3.4.2}$$

It is direct to verify that $\cos 3x$ is another solution to the S-L equation in (3.4.1) with $\lambda = 1$, and hence

$$\Phi(x,1) = \begin{pmatrix} \operatorname{sgn} x \sin 3x & -(\cos 3x)/3 \\ 3\cos 3x & \operatorname{sgn} x \sin 3x \end{pmatrix},$$

$$\Psi = \Phi(\pi/4, 1) = \frac{1}{\sqrt{2}}\begin{pmatrix} 1 & 1/3 \\ -3 & 1 \end{pmatrix}.$$

(3.4.3)

Note that exactly five zeros of $\cos 3x$ on $[-3\pi/2, \pi/4]$ have neighborhoods on which $f < 0$ a.e., and only one zero of $\cos 3x$ in $(-3\pi/2, \pi/4)$ has a neighborhood on which $f > 0$ a.e. Thus, by Remark 3.1.1, (3.4.3) and Corollary 4.11 in [98], the index of 1 as an eigenvalue for the BC S_{0,β_0} is -3. Here we remark that (1.28) in [98] and (3.1.10) use different multiples of π. Since

$$\beta_{0,K} = \operatorname{arccot}(-3\sqrt{3}) > \pi/2 > \beta_0 = \operatorname{arccot} 3, \qquad (3.4.4)$$

$\lambda_{-3} = 1$ by Theorem 3.1.1. In fact, the codes with psiYY= 2 obtain the following approximations of the eigenvalues on the interval [0,12] and their indices

$$\lambda_{-4} \approx 0.55262, \quad \lambda_{-3} \approx 1.0, \quad \lambda_{-2} \approx 1.52118,$$
$$\lambda_{-1} \approx 1.71224, \quad \lambda_0 \approx 2.80841, \quad \lambda_1 \approx 9.72016.$$

(3.4.5)

Example 3.4.2 Let $q(x) = x^2 - \operatorname{sgn} x$, and $w(x) = x^2$. Then, 1 is an eigenvalue of the S-L problem

$$-(fy')' + qy = \lambda wy \text{ in } (-3\pi/2, \pi/4), \qquad Y(\pi/4) = KY(-3\pi/2) \qquad (3.4.6)$$

with an eigenfunction $y(x) = \operatorname{sgn} x \sin x$, where

$$K = \frac{1}{\sqrt{2}}\begin{pmatrix} -1 & 1 \\ -1 & -1 \end{pmatrix}. \qquad (3.4.7)$$

It is direct to verify that $\cos x$ is another solution to the S-L equation in (3.4.6) with $\lambda = 1$, and hence

$$\Phi(x,1) = \begin{pmatrix} -\operatorname{sgn} x \sin x & \cos x \\ -\cos x & -\operatorname{sgn} x \sin x \end{pmatrix},$$

$$\Psi = \Phi(\pi/4, 1) = \frac{1}{\sqrt{2}}\begin{pmatrix} -1 & 1 \\ -1 & -1 \end{pmatrix}.$$

(3.4.8)

Note that exactly two zeros of $\cos t$ on $[-3\pi/2, \pi/4]$ have neighborhoods on which $f < 0$ a.e., and no zeros of $\cos x$ in $(-3\pi/2, \pi/4)$ have a neighborhood on which $f > 0$ a.e. Thus, the index of 1 as an eigenvalue for S_{0,β_0} is -1. Since

$$\beta_0 = \beta_{0,K} = 3\pi/4, \quad \beta_1 = \beta_{1,K} = \pi/4, \qquad (3.4.9)$$

$\lambda_{-1} = \lambda_0 = 1$ by Theorem 3.1.1. In fact, codes with psiYY= 3 yield the following approximations of the eigenvalues on the interval $[0, 6]$ and their indices

$$\lambda_{-3} \approx 0.14785, \quad \lambda_{-2} \approx 0.37465, \quad \lambda_{-1} = \lambda_0 \approx 1.0, \quad \lambda_1 \approx 2.23275. \quad (3.4.10)$$

Example 3.4.3 Let $q(x) = x^2 - 25\mathrm{sgn}x$, and $w(x) = x^2$. Then, 1 is an eigenvalue of the S-L problem

$$-(fy')' + qy = \lambda w y \text{ in } (-3\pi/2, \pi/4), \quad Y(\pi/4) = K Y(-3\pi/2) \quad (3.4.11)$$

with an eigenfunction $y(x) = \mathrm{sgn}x \sin 5x$, where

$$K = \frac{1}{\sqrt{2}} \begin{pmatrix} 1 & -2/5 \\ 5 & 0 \end{pmatrix}, \quad (3.4.12)$$

It is direct to verify that $\cos 5x$ is another solution to the S-L equation in (3.4.11) with $\lambda = 1$, and hence

$$\Phi(x, 1) = \begin{pmatrix} -\mathrm{sgn}x \sin 5x & (\cos 5x)/5 \\ -5 \cos 5x & -\mathrm{sgn}x \sin 5x \end{pmatrix},$$

$$\Psi = \Phi(\pi/4, 1) = \frac{1}{\sqrt{2}} \begin{pmatrix} 1 & -1/5 \\ 5 & 1 \end{pmatrix}. \quad (3.4.13)$$

Note that exactly eight zeros of $\cos 5x$ on $[-3\pi/2, \pi/4]$ have neighborhoods on which $f < 0$ a.e., and only one zero of $\cos 5x$ in $(-3\pi/2, \pi/4)$ has a neighborhood on which $f > 0$ a.e. Thus, the index of 1 as an eigenvalue for $\mathbf{S}_{0,\beta_0}$ is -6. Since

$$\beta_0 = \mathrm{arccot}(-5) > \pi/2 > \beta_{0,K} = \pi/2. \quad (3.4.14)$$

Thus, $\lambda_{-5} = 1$ by Theorem 3.1.1. Actually, codes with psiYY=4 obtain the following approximations of the eigenvalues on the interval $[-3, 3]$ and their indices

$$\lambda_{-8} \approx -1.69784, \quad \lambda_{-7} \approx -1.17353, \quad \lambda_{-6} \approx 0.71166,$$
$$\lambda_{-5} \approx 0.99999, \quad \lambda_{-4} \approx 2.21812, \quad \lambda_{-3} \approx 2.44785. \quad (3.4.15)$$

For the rest of this section, let $q(x) = 1 \cdot \mathrm{sgn}x$. Assume that $a < 0$ and $b > 0$ satisfy $a + b \in (0, \pi)$. Then, for $\lambda \neq 0$ and $\lambda \neq 2$, the fundamental solution matrix of the S-L equation $-(fy')' + qy = \lambda y$ in (a, b) is

$$\Phi(x, \lambda) = \begin{cases} \Phi_L(x, \lambda) & \text{for } x \leqslant 0, \\ \Phi_R(x, \lambda)\Phi_L(0, \lambda) & \text{for } x > 0, \end{cases} \quad (3.4.16)$$

where

$$\Phi_L(x,\lambda) = \begin{pmatrix} \cos[\sqrt{2-\lambda}\,(x-a)] & -\dfrac{1}{\sqrt{2-\lambda}}\sin[\sqrt{2-\lambda}\,(x-a)] \\ \sqrt{2-\lambda}\,\sin[\sqrt{2-\lambda}\,(x-a)] & \cos[\sqrt{2-\lambda}\,(x-a)] \end{pmatrix},$$

$$\Phi_R(x,\lambda) = \begin{pmatrix} \cos(\sqrt{\lambda}\,x) & \dfrac{1}{\sqrt{\lambda}}\sin(\sqrt{\lambda}\,x) \\ -\sqrt{\lambda}\,\sin(\sqrt{\lambda}\,x) & \cos(\sqrt{\lambda}\,x) \end{pmatrix}.$$

$$(3.4.17)$$

Thus, for the special value $\lambda_* = 1$ of λ,

$$\phi_{12}(x,1) = \begin{cases} -\sin(x-a) & \text{for } x \leqslant 0, \\ \sin(x+a) & \text{for } x > 0, \end{cases} \qquad (3.4.18)$$

$$\Psi = \Phi(b,1) = \begin{pmatrix} \cos(a+b) & \sin(a+b) \\ -\sin(a+b) & \cos(a+b) \end{pmatrix},$$

and hence $\beta_0 = a+b$, while $\beta_1 = a+b+\pi/2$ if $a+b \in (0,\pi/2]$ and $a+b-\pi/2$ if $a+b \in (\pi/2,\pi)$. The next six examples use these observations.

Example 3.4.4 Consider the S-L problem

$$-(fy')' + qy = \lambda y \text{ in } (-\pi/2, 3\pi/4),$$

$$Y(3\pi/4) = \frac{1}{\sqrt{2}}\begin{pmatrix} 0 & 1 \\ -2 & 1 \end{pmatrix} Y(-\pi/2). \qquad (3.4.19)$$

Then, 1 is an eigenvalue by Theorem 3.1.1 and (3.4.18). Note that exactly one zero of

$$\phi_{12}(x,1) = \begin{cases} -\sin(x+\pi/2) = -\cos x & \text{for } x \leqslant 0, \\ \sin(x-\pi/2) = -\cos x & \text{for } x > 0 \end{cases} \qquad (3.4.20)$$

on $[-\pi/2, 3\pi/4]$ has a neighborhood on which $f < 0$ a.e., and only one zero of $\phi_{12}(x,1)$ in $(-\pi/2, 3\pi/4)$ has a neighborhood on which $f > 0$ a.e. Thus, the index of 1 as an eigenvalue for $\mathbf{S}_{0,\beta_0}$ is 1. In this example, $\lambda_* = 1.0$ is used by codes, and hence

$$\beta_{1,K} = \pi > \beta_1 = 3\pi/4 > \beta_{0,K} = \beta_0 = \pi/4. \qquad (3.4.21)$$

Thus, $\lambda_1 = 1$ by Theorem 3.1.1. Actually, codes yield the following approximations of the eigenvalues on the interval $[-20, 2]$

$$\lambda_{-1} \approx -11.4566, \quad \lambda_0 \approx -2.86041, \quad \lambda_1 \approx 1.0, \quad \lambda_2 \approx 1.63993. \qquad (3.4.22)$$

Example 3.4.5 Take the S-L problem

$$-(fy')' + qy = \lambda y \text{ in } (-\pi/2, 3\pi/4), \qquad Y(3\pi/4) = KY(-\pi/2), \qquad (3.4.23)$$

where
$$K = \begin{pmatrix} \sqrt{2}/(1-\sqrt{3}) & 1/\sqrt{2} \\ \sqrt{6}/(1-\sqrt{3}) & 1/\sqrt{2} \end{pmatrix}. \tag{3.4.24}$$

Then, 1 is an eigenvalue. Note that $\phi_{12}(x,1)$ is the same as in (3.4.20). Thus, the index of 1 as an eigenvalue for $\mathbf{S}_{0,\beta_0}$ is 1. In this example, $\lambda_* = 1.0$ is used by the codes, and hence

$$\beta_1 = 3\pi/4 > \beta_{0,K} = \beta_0 = \pi/4 > \beta_{1,K} = \pi/6. \tag{3.4.25}$$

Thus, $\lambda_1 = 1$ by Theorem 3.1.1. In fact, codes obtain the following approximations of the eigenvalues on the interval $[-30,10]$:

$$\begin{aligned} &\lambda_{-2} \approx -25.6177, \quad \lambda_{-1} \approx -8.41435, \quad \lambda_0 \approx -2.24248, \\ &\lambda_1 \approx 1.0, \quad \lambda_2 \approx 3.76805, \quad \lambda_3 \approx 9.40145. \end{aligned} \tag{3.4.26}$$

Example 3.4.6 Pick the S-L problem

$$-(fy')' + qy = \lambda y \text{ in } (-\pi/4, \pi), \quad Y(\pi) = \frac{1}{\sqrt{2}} \begin{pmatrix} -2 & 1 \\ 0 & -1 \end{pmatrix} Y(-\pi/4). \tag{3.4.27}$$

Then, 1 is an eigenvalue. Note that exactly one zero of

$$\phi_{12}(x,1) = \begin{cases} -\sin(x+\pi/4) & \text{for } x \leqslant 0, \\ \sin(x-\pi/4) & \text{for } x > 0 \end{cases} \tag{3.4.28}$$

on $[-\pi/4, \pi]$ has a neighborhood on which $f < 0$ a.e., and only one zero of $\phi_{12}(x,1)$ in $(-\pi/4, \pi)$ has a neighborhood on which $f > 0$ a.e. Thus, the index of 1 as an eigenvalue for $\mathbf{S}_{0,\beta_0}$ is 1. In this example, $\lambda_* = 1.0$ is used by the codes, and hence

$$\beta_{0,K} = \beta_0 = 3\pi/4 > \beta_{1,K} = \pi/2 > \beta_1 = \pi/4. \tag{3.4.29}$$

Thus, $\lambda_1 = 1$ by Theorem 3.1.1. Actually, the codes obtain the following approximations of the eigenvalues on the interval $[-50,4]$

$$\lambda_{-1} \approx -42.7771, \quad \lambda_0 \approx -4.22493, \quad \lambda_1 \approx 1.0, \quad \lambda_2 \approx 1.55760. \tag{3.4.30}$$

Example 3.4.7 Consider the S-L problem

$$-(fy')' + qy = \lambda y \text{ in } (-\pi/2, 3\pi/4), \quad Y(3\pi/4) = \frac{1}{\sqrt{2}} \begin{pmatrix} 2 & 1 \\ 0 & 1 \end{pmatrix} Y(-\pi/2). \tag{3.4.31}$$

Then, 1 is an eigenvalue. Note that $\phi_{12}(x,1)$ is the same as in (3.4.20). Thus, the index of

1 as an eigenvalue for S_{0,β_0} is 1. In this example, $\lambda_* = 0.99999$ is used by the codes, and hence

$$\beta_1 = 3\pi/4 > \beta_{1,K} = \pi/2 > \beta_{0,K} = \beta_0 = \pi/4. \quad (3.4.32)$$

Thus, $\lambda_2 = 1$ by Theorem 3.1.1. In fact, the codes with psiYY= 4 yield the following approximations of the eigenvalues on the interval $[-20, 20]$

$$\lambda_{-1} \approx -13.3089, \quad \lambda_0 \approx -3.25674, \quad \lambda_1 \approx 0.55946,$$
$$\lambda_2 \approx 0.99999, \quad \lambda_3 \approx 5.36909, \quad \lambda_4 \approx 13.0323. \quad (3.4.33)$$

Example 3.4.8 Take the S-L problem

$$-(fy')' + qy = \lambda y \text{ in } (-\pi/4, \pi), \quad Y(\pi) = \frac{1}{\sqrt{2}} \begin{pmatrix} 0 & 1 \\ -2 & -1 \end{pmatrix} Y(-\pi/4). \quad (3.4.34)$$

Then, 1 is an eigenvalue. Note that $\phi_{12}(x, 1)$ is the same as in (3.4.20). Thus, the index of 1 as an eigenvalue for S_{0,β_0} is 1. In this example, $\lambda_* = 0.99999$ is used by the codes, and hence

$$\beta_{1,K} = \pi > \beta_{0,K} = \beta_0 = 3\pi/4 > \beta_1 = \pi/4. \quad (3.4.35)$$

Thus, $\lambda_2 = 1$ by Theorem 3.1.1. Actually, codes with psiYY= 4 obtain the following approximations of the eigenvalues on the interval $[-50, 10]$

$$\lambda_{-1} \approx -47.8246, \quad \lambda_0 \approx -8.64548, \quad \lambda_1 \approx 0.63489,$$
$$\lambda_2 \approx 0.99999, \quad \lambda_3 \approx 4.73269, \quad \lambda_4 \approx 8.39566. \quad (3.4.36)$$

Example 3.4.9 Pick the S-L problem

$$-(fy')' + qy = \lambda y \text{ in } (-\pi/4, \pi), \quad Y(\pi) = KY(-\pi/4), \quad (3.4.37)$$

where

$$K = \begin{pmatrix} -\sqrt{2}/(1+\sqrt{3}) & 1/\sqrt{2} \\ -\sqrt{6}/(1+\sqrt{3}) & -1/\sqrt{2} \end{pmatrix}. \quad (3.4.38)$$

Then, 1 is an eigenvalue. Note that $\phi_{12}(x, 1)$ is the same as in (3.4.28). Thus, the index of 1 as an eigenvalue for S_{0,β_0} is 1. In this example, $\lambda_* = 0.99999$ is used by the codes, and hence

$$\beta_{0,K} = \beta_0 = 3\pi/4 > \beta_1 = \pi/4 > \beta_{1,K} = \pi/6. \quad (3.4.39)$$

Thus, $\lambda_2 = 1$ by Theorem 3.1.1. In fact, the codes with psiYY= 4 obtain the following approximations of the eigenvalues on the interval $[-50, 10]$

$$\lambda_{-1} \approx -45.9756, \quad \lambda_0 \approx -7.01204, \quad \lambda_1 \approx 0.88427,$$
$$\lambda_2 \approx 0.99999, \quad \lambda_3 \approx 5.08047, \quad \lambda_4 \approx 8.44679. \tag{3.4.40}$$

3.5 Examples about $\beta_*(\alpha)$ and $\beta(\alpha)$

Example 3.5.1 Let $q(x) = x - 1$, and $w(x) = x$. Then, 1 is an eigenvalue of the S-L problem

$$-y'' + qy = \lambda w y \text{ in } (0, \pi), \quad \boldsymbol{Y}(\pi) = \begin{pmatrix} -1 & -1 \\ 0 & -1 \end{pmatrix} \boldsymbol{Y}(0) \tag{3.5.1}$$

with an eigenfunction $y(x) = \cos x$. It is direct to verify that $\sin x$ is another solution to the S-L equation (3.5.1) with $\lambda = 1$. The index of 1 as an eigenvalue for $\mathbf{S}_{0,\beta_0}$ is 1, and have the simple eigenvalues

$$-7.93477, \quad 1.0, \quad 1.57704, \quad 6.69360, \quad 9.47920 \tag{3.5.2}$$

on the interval $[-10, 10]$.

Figure 3.5.1 and Figure 3.5.2 below illustrate the functions $\beta_*(\alpha)$ and $\beta(\alpha)$, respectively, where $\alpha_{0,*} \approx 9.09083 \times 10^{-8}$, $\alpha_0 \approx 2.35619$ are obtained by the code.

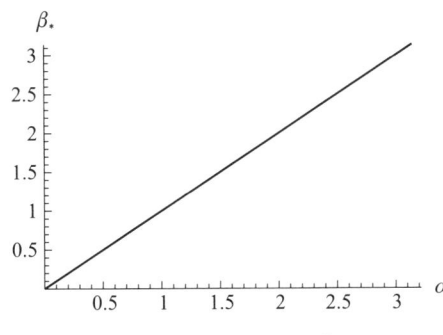

Figure 3.5.1 Graph of $\beta_*(\alpha)$

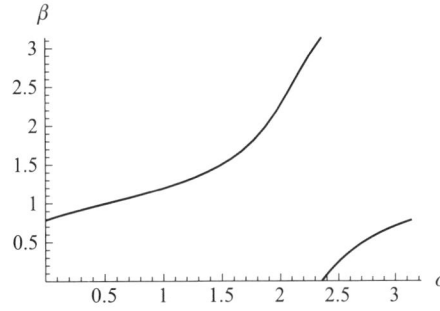

Figure 3.5.2 Graph of $\beta(\alpha)$

The codes choose $\alpha_1 \approx 1.17807$ and $\alpha_2 \approx 2.74876$, then give

$$1.17807 \approx \beta_{1,*} < \beta_1 \approx 1.28586, \quad 2.74876 \approx \beta_{2,*} > \beta_2 \approx 0.529788. \tag{3.5.3}$$

$\lambda_2 = 1$ by Theorem 3.1.1. In fact, the codes give their indices

$$\lambda_1 \approx -7.93477, \lambda_2 \approx 1.0, \lambda_3 \approx 1.57704, \lambda_4 \approx 6.69360, \lambda_5 \approx 9.47920. \tag{3.5.4}$$

Example 3.5.2 Let $q(x) = x - 1$, and $w(x) = x$. Then, 1 is an eigenvalue of the S-L problem

$$-y'' + qy = \lambda wy \text{ in } (0, 7\pi/4), \qquad Y(7\pi/4) = \frac{1}{\sqrt{2}}\begin{pmatrix} 1 & -2 \\ 1 & 0 \end{pmatrix} Y(0) \qquad (3.5.5)$$

with an eigenfunction $y(x) = \cos x$. It is direct to verify that $\sin x$ is another solution to the S-L equation in (3.5.5) with $\lambda = 1$. The index of 1 as an eigenvalue for S_{0,β_0} is 2, and have the simple eigenvalues

$$0.24995, \quad 0.90578, \quad 0.99999, \quad 2.04145, \quad 2.61244 \qquad (3.5.6)$$

on the interval $[-4,4]$.

Figure 3.5.3 and Figure 3.5.4 below illustrate the functions $\beta_*(\alpha)$ and $\beta(\alpha)$, respectively, where $\alpha_{0,*} \approx 0.785398$, $\alpha_0 \approx 1.107148$.

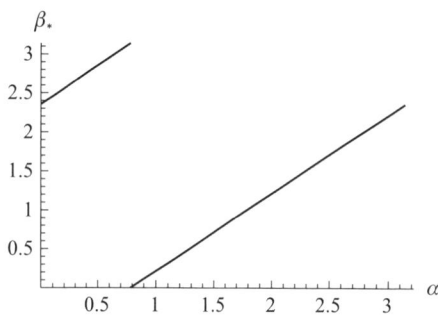

Figure 3.5.3 Graph of $\beta_*(\alpha)$

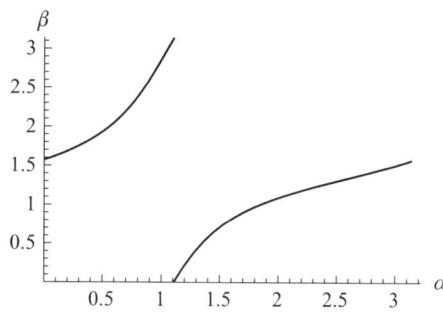

Figure 3.5.4 Graph of $\beta(\alpha)$

The codes choose $\alpha_1 \approx 2.51693$ and $\alpha_2 \approx 0.946261$, then give

$$1.73153 \approx \beta_{1,*} > \beta_1 \approx 1.31178, \quad 0.160862 \approx \beta_{2,*} < \beta_2 \approx 2.72578. \qquad (3.5.7)$$

Thus, $\lambda_3 = 1$ by Theorem 3.1.1. Actually, the codes obtain their indices

$$\lambda_1 \approx 0.24995, \lambda_2 \approx 0.90578, \lambda_3 \approx 0.99999, \lambda_4 \approx 2.04145, \lambda_5 \approx 2.61244. \qquad (3.5.8)$$

If the BC chosen is

$$Y(7\pi/4) = \begin{pmatrix} 1/\sqrt{2} & -\sqrt{2}/(1+\sqrt{3}) \\ 1/\sqrt{2} & \sqrt{6}/(1+\sqrt{3}) \end{pmatrix} Y(0), \qquad (3.5.9)$$

then, the simple eigenvalues are as follows

$$0.28841, \quad 1.0, \quad 1.03071, \quad 2.30948, \quad 2.61988, \quad 4.63613 \qquad (3.5.10)$$

on the interval $[-5, 5]$.

Figure 3.5.5 and Figure 3.5.6 below illustrate the functions $\beta_*(\alpha)$ and $\beta(\alpha)$, respectively, where $\alpha_{0,*} \approx 0.785398$, $\alpha_0 \approx 0.631914$.

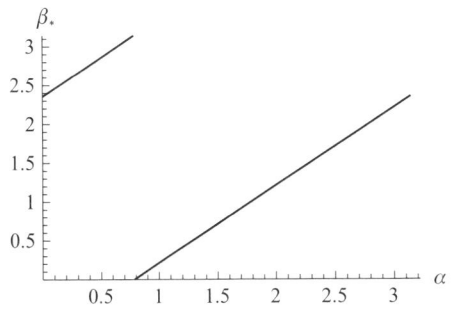

Figure 3.5.5 Graph of $\beta_*(\alpha)$

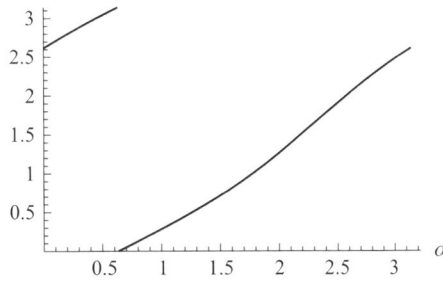

Figure 3.5.6 Graph of $\beta(\alpha)$

The codes choose $\alpha_1 \approx 2.2793$ and $\alpha_2 \approx 0.708564$, then give

$$1.4939 \approx \beta_{1,*} < \beta_1 \approx 1.62393, \quad 3.06476 \approx \beta_{2,*} > \beta_2 \approx 0.0587491, \tag{3.5.11}$$

thus $\lambda_2 = 1$. In fact, the codes yield their indices

$$\lambda_1 \approx 0.28841, \ \lambda_2 \approx 1.0, \ \lambda_3 \approx 1.03071, \ \lambda_4 \approx 2.30948, \ \lambda_5 \approx 2.61988, \ \lambda_6 \approx 4.63613. \tag{3.5.12}$$

If we change the BC to

$$\boldsymbol{Y}(7\pi/4) = \frac{1}{\sqrt{2}} \begin{pmatrix} 1 & -1 \\ 1 & 1 \end{pmatrix} \boldsymbol{Y}(0), \tag{3.5.13}$$

then, the eigenvalues and their multiplicities are as follows

$$1, 0.27973, \quad 2, 0.99985, \quad 1, 2.23403, \quad 1, 2.61682, \quad 1, 4.52107 \tag{3.5.14}$$

on the interval $[-5, 5]$.

Figure 3.5.7 and Figure 3.5.8 below illustrate the functions $\beta_*(\alpha)$ and $\beta(\alpha)$, respectively, where $\alpha_{0,*} \approx 0.785398$, $\alpha_0 \approx 0.785398$.

The codes choose $\alpha_1 \approx 0.156848$ and $\alpha_2 \approx 0.392437$, then give

$$\beta_{1,*} = \beta_1 \approx 2.51304, \quad \beta_{2,*} = \beta_2 \approx 1.57069, \tag{3.5.15}$$

thus, $\lambda_2 = \lambda_3 = 1$. Actually, the codes obtain their indices

$$\lambda_1 \approx 0.27973, \quad \lambda_2 = \lambda_3 \approx 0.99985,$$

$$\lambda_4 \approx 2.23403, \quad \lambda_5 \approx 2.61682, \quad \lambda_6 \approx 4.52107. \tag{3.5.16}$$

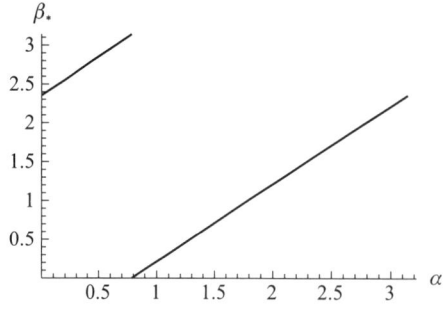

Figure 3.5.7 Graph of $\beta_*(\alpha)$

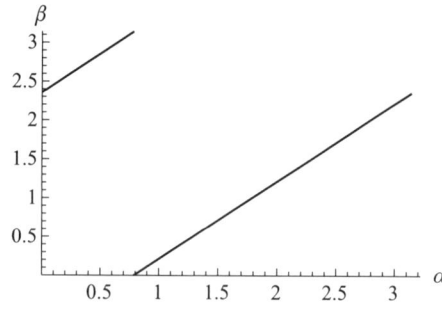

Figure 3.5.8 Graph of $\beta(\alpha)$

In the rest of this section, we always set $f(x) = \mathrm{sgn}\,x$.

Example 3.5.3 Let $q(x) = x^2 - 9\,\mathrm{sgn}\,x$, and $w(x) = x^2$. Then, 1 is an eigenvalue of the S-L problem

$$-(fy')' + qy = \lambda w y \text{ in } (-3\pi/2, \pi/4), \qquad Y(\pi/4) = \boldsymbol{K}\,Y(-3\pi/2) \qquad (3.5.17)$$

with an eigenfunction $y(x) = \mathrm{sgn}\,x \sin 3x$, where

$$\boldsymbol{K} = \sqrt{2}\begin{pmatrix} 1/2 & -1/[3(\sqrt{3}-1)] \\ -3/2 & \sqrt{3}/(\sqrt{3}-1) \end{pmatrix}. \qquad (3.5.18)$$

It is direct to verify that $\cos 3x$ is another solution to the S-L equation in (3.5.17) with $\lambda = 1$. The index of 1 as an eigenvalue for $\mathbf{S}_{0,\beta_0}$ is -3, and the simple eigenvalues are as follows

$$0.55262, \quad 1.0, \quad 1.52118, \quad 1.71224, \quad 2.80841, \quad 9.72016 \qquad (3.5.19)$$

on the interval $[0, 12]$.

Figure 3.5.9 and Figure 3.5.10 below illustrate the function $\beta_*(\alpha)$ and $\beta(\alpha)$, where $\alpha_{0,*} \approx 1.31345$, $\alpha_0 \approx 0.738686$.

The codes choose $\alpha_1 \approx 2.5973$ and $\alpha_2 \approx 1.02423$, then give

$$0.151065 \approx \beta_{1,*} < \beta_1 \approx 2.91408, \quad 3.11506 \approx \beta_{2,*} > \beta_2 \approx 1.83582. \qquad (3.5.20)$$

Thus, $\lambda_{-3} = 1$ by Theorem 3.1.1. In fact, the codes obtain their indices

$$\lambda_{-4} \approx 0.55262, \ \lambda_{-3} \approx 1.0, \ \lambda_{-2} \approx 1.52118, \ \lambda_{-1} \approx 1.71224,$$

$$\lambda_0 \approx 2.80841, \ \lambda_1 \approx 9.72016. \qquad (3.5.21)$$

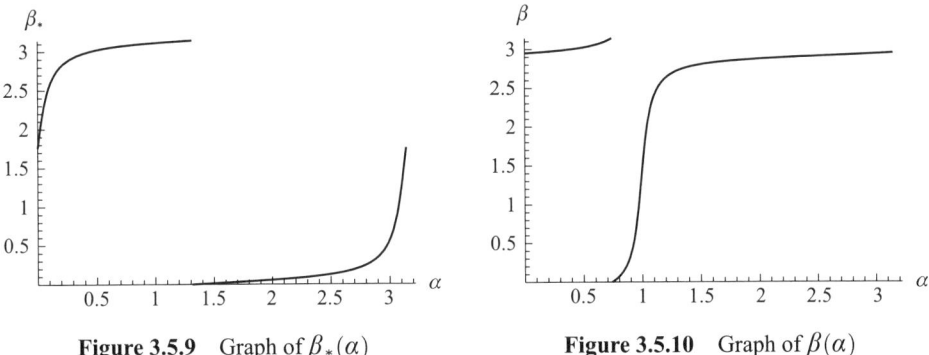

Figure 3.5.9 Graph of $\beta_*(\alpha)$ **Figure 3.5.10** Graph of $\beta(\alpha)$

Example 3.5.4 Let $q(x) = x^2 - \mathrm{sgn}x$, and $w(x) = x^2$. Then, 1 is an eigenvalue of the S-L problem

$$-(fy')' + qy = \lambda wy \text{ in } (-3\pi/2, \pi/4), \quad Y(\pi/4) = KY(-3\pi/2) \quad (3.5.22)$$

with an eigenfunction $y(x) = \mathrm{sgn}x \sin x$, where

$$K = \frac{1}{\sqrt{2}} \begin{pmatrix} -1 & 1 \\ -1 & -1 \end{pmatrix}. \quad (3.5.23)$$

It is direct to verify that $\cos x$ is another solution to the S-L equation in (3.5.22) with $\lambda = 1$. The index of 1 as an eigenvalue for $\mathbf{S}_{0,\beta_0}$ is -1, and the eigenvalues and their multiplicities are as follows

$$1, 0.14785, \quad 1, 0.37465, \quad 2, 1.0, \quad 1, 2.23275 \quad (3.5.24)$$

on the interval $[0, 6]$ by [5].

Figure 3.5.11 and Figure 3.5.12 below illustrate the functions $\beta_*(\alpha)$ and $\beta(\alpha)$, where $\alpha_{0,*} \approx 0.785398$, $\alpha_0 \approx 0.785398$.

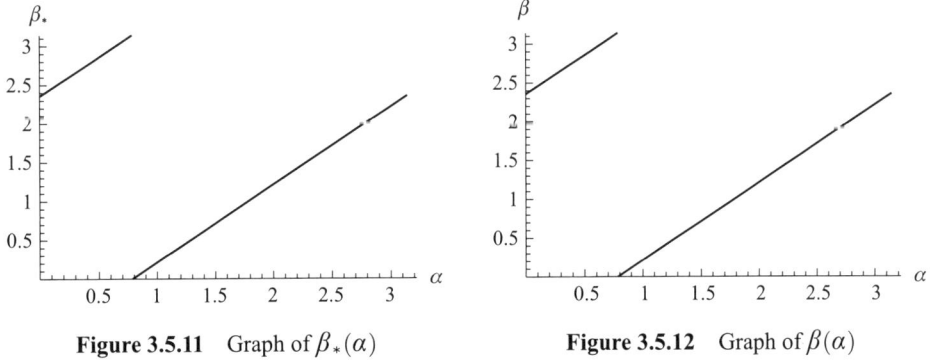

Figure 3.5.11 Graph of $\beta_*(\alpha)$ **Figure 3.5.12** Graph of $\beta(\alpha)$

The codes choose $\alpha_1 \approx 2.63304$ and $\alpha_2 \approx 1.06145$, then give

$$\beta_{1,*} = \beta_1 \approx 2.51304, \quad \beta_{2,*} = \beta_2 \approx 1.57069. \tag{3.5.25}$$

Thus, $\lambda_{-1} = \lambda_0 = 1$ by Theorem 3.1.1. In fact, the codes yield their indices

$$\lambda_{-3} \approx 0.14785, \quad \lambda_{-2} \approx 0.37465, \quad \lambda_{-1} = \lambda_0 \approx 1.0, \quad \lambda_1 \approx 2.23275. \tag{3.5.26}$$

Example 3.5.5 Let $q(x) = x^2 - 25\operatorname{sgn} x$, and $w(x) = x^2$. Then, 1 is an eigenvalue of the S-L problem

$$-(fy')' + qy = \lambda wy \text{ in } (-3\pi/2, \pi/4), \quad Y(\pi/4) = K\,Y(-3\pi/2) \tag{3.5.27}$$

with an eigenfunction $y(x) = \operatorname{sgn} x \sin 5x$, where

$$K = \frac{1}{\sqrt{2}} \begin{pmatrix} 1 & -2/5 \\ 5 & 0 \end{pmatrix}. \tag{3.5.28}$$

It is direct to verify that $\cos 5x$ is another solution to the S-L equation in (3.5.27) with $\lambda = 1$. The index of 1 as an eigenvalue for $\mathbf{S}_{0,\beta_0}$ is -6 and has the eigenvalues

$$-1.69784, \quad -1.17353, \quad 0.71166, \quad 0.99999, \quad 2.21812, \quad 2.44785 \tag{3.5.29}$$

on the interval $[-3, 3]$.

Figure 3.5.13 and Figure 3.5.14 below illustrate the functions $\beta_*(\alpha)$ and $\beta(\alpha)$, where $\alpha_{0,*} \approx 0.197395$, $\alpha_0 \approx 0.380506$.

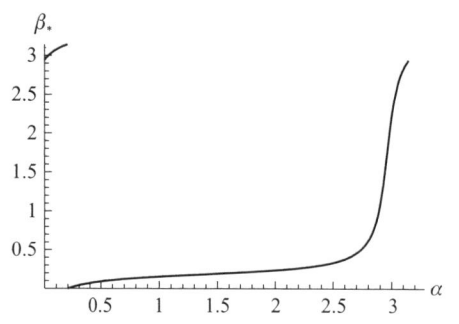

Figure 3.5.13 Graph of $\beta_*(\alpha)$

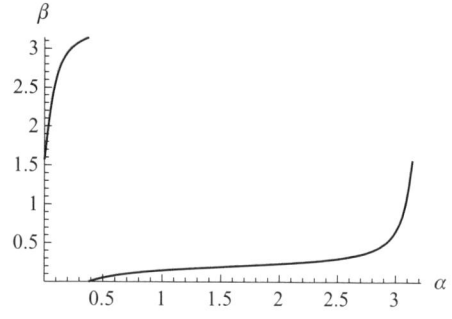

Figure 3.5.14 Graph of $\beta(\alpha)$

The codes choose $\alpha_1 \approx 1.85953$ and $\alpha_2 \approx 0.288777$, then give

$$0.221566 \approx \beta_{1,*} > \beta_1 \approx 0.220137, \quad 0.0390406 \approx \beta_{2,*} < \beta_2 \approx 3.07242. \tag{3.5.30}$$

Thus, $\lambda_{-5} = 1$ by Theorem 3.1.1. Actually, the codes find their indices

$$\lambda_{-8} \approx -1.69784, \quad \lambda_{-7} \approx -1.17353, \quad \lambda_{-6} \approx 0.71166, \tag{3.5.31}$$

$$\lambda_{-5} \approx 0.99999, \quad \lambda_{-4} \approx 2.21812, \quad \lambda_{-3} \approx 2.44785.$$

Chapter 4 Relations among eigenvalues of Sturm-Liouville problems

Consider the S-L problem consisting of a regular S-L equation

$$-(fy')' + qy = \lambda wy \text{ in } (a,b) \tag{4.0.1}$$

and a separable self-adjoint boundary condition (BC)

$$\cos\alpha \cdot y(a) - \sin\alpha \cdot (fy')(a) = 0 = \cos\beta \cdot y(b) - \sin\beta \cdot (fy')(b), \tag{4.0.2}$$

where

$$-\infty \leqslant a < b \leqslant +\infty, \quad 1/f, q, w \in L((a,b), \mathbb{R}), \quad w > 0 \text{ a.e. in } (a,b), \tag{4.0.3}$$

$$\operatorname{sgn}(x-c)f(x) > 0 \text{ a.e. in } (a,b) \text{ for some } c \in (a,b), \tag{4.0.4}$$

$$\alpha \in (0, \pi], \quad \beta \in (0, \pi], \tag{4.0.5}$$

and $\lambda \in \mathbb{C}$ is the so-called spectral parameter. Note that the leading coefficient function f changes the sign exactly once, i.e., at c. We want to relate the eigenvalues

$$\cdots < \lambda_{-2} < \lambda_{-1} < \lambda_0 < \lambda_1 < \lambda_2 < \cdots \tag{4.0.6}$$

of this problem to the eigenvalues

$$\mu_0(\gamma) > \mu_{-1}(\gamma) > \mu_{-2}(\gamma) > \cdots \longrightarrow -\infty \tag{4.0.7}$$

and

$$\nu_0(\gamma) < \nu_1(\gamma) < \nu_2(\gamma) < \cdots \longrightarrow +\infty \tag{4.0.8}$$

of two corresponding one-parameter families of problems

$$-(fy')' + qy = \lambda wy \text{ in } (a,c), \tag{4.0.9}$$

$$\cos\alpha \cdot y(a) - \sin\alpha \cdot (fy')(a) = 0 = \cos\gamma \cdot y(c) - \sin\gamma \cdot (fy')(c) \tag{4.0.10}$$

and

$$-(fy')' + qy = \lambda wy \text{ in } (c,b), \tag{4.0.11}$$

$$\cos\gamma \cdot y(c) - \sin\gamma \cdot (fy')(c) = 0 = \cos\beta \cdot y(b) - \sin\beta \cdot (fy')(b) \tag{4.0.12}$$

with a definite f (i.e., f does not change the sign in each of the two subintervals above), where the parameter $\gamma \in [0, \pi)$. Here the indices of the eigenvalues in (4.0.7) and (4.0.8) are given according to the Prüfer angle characterization of eigenvalues. This explains why in (4.0.5) we do not take $[0, \pi)$ as the range of α, see Lemma 4.1.4 below for details.

4.1 Notation and basic results

The following fact about the reality of eigenvalues is well-known [125].

Lemma 4.1.1 *The eigenvalues of the S-L problem consisting of (4.0.1) and (4.0.2) are all real.*

A proof of the next basic lemma can be found in [33].

Lemma 4.1.2 *Each μ_j is continuous and strictly increasing on $[0, \pi)$, and*

$$\lim_{\gamma \to \pi^-} \mu_0(\gamma) = +\infty, \qquad \lim_{\gamma \to \pi^-} \mu_{-j}(\gamma) = \mu_{-j+1}(0) \text{ for } j \geq 1, \qquad (4.1.1)$$

$$\lim_{j \to +\infty} \mu_{-j}(\gamma) = -\infty \text{ for } \gamma \in [0, \pi); \qquad (4.1.2)$$

while each ν_j is continuous and strictly decreasing on $[0, \pi)$, and

$$\lim_{\gamma \to \pi^-} \nu_0(\gamma) = -\infty, \qquad \lim_{\gamma \to \pi^-} \nu_j(\gamma) = \nu_{j-1}(0) \text{ for } j \geq 1, \qquad (4.1.3)$$

$$\lim_{j \to +\infty} \nu_j(\gamma) = +\infty \text{ for } \gamma \in [0, \pi). \qquad (4.1.4)$$

For any non-trivial real solution y to (4.0.1), there are two unique absolutely continuous real functions ρ and θ on $[a, b]$ such that $\rho(x, \lambda) \neq 0$ for all $x \in [a, b]$, and

$$y = \rho \sin \theta, \quad fy' = \rho \cos \theta, \quad 0 \leq \theta(a, \lambda) < \pi. \qquad (4.1.5)$$

The function θ is called the Prüfer angle of the solution y. Note that y satisfies the self-adjoint BC (4.0.2) if and only if

$$\theta(a, \lambda) = \alpha, \qquad \theta(b, \lambda) = \beta + n\pi \text{ for some } n \in \mathbb{Z}. \qquad (4.1.6)$$

Since each eigenvalue has geometric multiplicity 1, all its real eigenfunctions share the same Prüfer angle, to be called the Prüfer angle of the eigenvalue. The differential equation for the Prüfer angle is

$$\theta' = \frac{1}{f} \cos^2 \theta + (\lambda w - q) \sin^2 \theta \text{ in } (a, b). \qquad (4.1.7)$$

The integrability conditions in (4.0.3) imply that for each $\lambda \in \mathbb{R}$, every initial condition $\theta(a, \lambda) = \alpha$ determines a unique solution $\theta(\cdot, \lambda)$ to (4.1.7). These solutions θ are called the Prüfer angles of (4.0.1).

For each $\alpha \in \mathbb{R}$, we have the following fact about the corresponding Prüfer angle $\theta = \theta(\cdot, \lambda)$ to (4.0.1), whose proof can be found in [125] and [126].

Lemma 4.1.3 If $x_* \in (a, b]$, then $\theta(x_*, \lambda)$ is strictly increasing in λ on $\mathbb{R}$.

The following lemma gives the Prüfer angle characterization of $\mu_{-j}(\gamma)$ and $\nu_k(\gamma)$.

Lemma 4.1.4 Let $\gamma \in [0, \pi)$.

(1) For each integer $j \geqslant 0$, $\mu_{-j}(\gamma)$ is the unique eigenvalue of the S-L problem consisting of (4.0.9) and (4.0.10) whose Prüfer angle $\theta(\cdot, \mu_{-j}(\gamma))$ satisfies

$$\theta(a, \mu_{-j}(\gamma)) = \alpha, \qquad \theta(c, \mu_{-j}(\gamma)) = \gamma - j\pi. \tag{4.1.8}$$

(2) For every integer $k \geqslant 0$, $\nu_k(\gamma)$ is the unique eigenvalue of the S-L problem consisting of (4.0.11) and (4.0.12) whose Prüfer angle $\theta(\cdot, \nu_k(\gamma))$ satisfies

$$\theta(c, \nu_k(\gamma)) = \gamma, \qquad \theta(b, \nu_k(\gamma)) = \beta + k\pi. \tag{4.1.9}$$

Proof. (2) Note that $\gamma \in [0, \pi)$ and $\beta \in (0, \pi]$. Thus, this part is well-known [126].

(1) Note that $-f > 0$ a.e. in (a, c). Let $\tilde{\alpha} = \pi - \alpha$ and $\tilde{\gamma} = \pi - \gamma$, then $\tilde{\alpha} \in [0, \pi)$, $\tilde{\gamma} \in (0, \pi]$, and the eigenvalues of the S-L problem

$$-(-fy')' + (-q)y = \lambda w y \quad \text{in } (a, c), \tag{4.1.10}$$

$$\cos \tilde{\alpha} \cdot y(a) - \sin \tilde{\alpha} \cdot (-fy')(a) = 0 = \cos \tilde{\gamma} \cdot y(c) - \sin \tilde{\gamma} \cdot (-fy')(c) \tag{4.1.11}$$

are $-\mu_0(\gamma), -\mu_{-1}(\gamma), -\mu_{-2}(\gamma), \cdots$. From (4.1.7) we deduce that if $\theta(x, \lambda)$ is the Prüfer angle of (4.0.9) fulfilling $\theta(a, \lambda) = \alpha$, then the Prüfer angle of (4.1.10) satisfying $\tilde{\theta}(a, \lambda) = \tilde{\alpha}$ is given by $\tilde{\theta}(x, \lambda) = \pi - \theta(x, -\lambda)$. Therefore, this part is a consequence of (2). $\square$

4.2 Geometric characterization of λ_n

First, we give the key lemma of this book.

Lemma 4.2.1 A real number $\lambda_\#$ is an eigenvalue of the S-L problem consisting of (4.0.1) and (4.0.2) if and only if there is a number $\gamma \in [0, \pi)$ such that $\lambda_\#$ is a common eigenvalue of the problem consisting of (4.0.9) and (4.0.10) and the problem consisting of (4.0.11) and (4.0.12).

In this case, an absolutely continuous function y in (a, b) with a continuous fy' is an eigenfunction for $\lambda_\#$ as an eigenvalue of the problem consisting of (4.0.1) and (4.0.2) if and only if $y|_{(a,c)}$ and $y|_{(c,b)}$ are eigenfunctions for $\lambda_\#$ as eigenvalues of the problem consisting of (4.0.9) and (4.0.10) and the problem consisting of of (4.0.11) and (4.0.12), respectively.

Proof. Assume that $\lambda_\# \in \mathbb{R}$ is an eigenvalue of the S-L problem consisting of (4.0.1) and (4.0.2). Then, there is a real-valued eigenfunction $y_\#$ for $\lambda_\#$. Let $\gamma \in [0, \pi)$ be the unique number such that $\cos\gamma \cdot y_\#(c) = \sin\gamma \cdot (fy'_\#)(c)$. Then, $\lambda_\#$ is a common eigenvalue of the problem consisting of (4.0.9) and (4.0.10), and the problem consisting of (4.0.11) and (4.0.12), with eigenfunctions $y_\#|_{(a,c)}$ and $y_\#|_{(c,b)}$, respectively.

Conversely, let $\lambda_\#$ be a common eigenvalue of the problem consisting of (4.0.9) and (4.0.10), and the problem consisting of (4.0.11) and (4.0.12) with real-valued eigenfunctions y_a and y_b, respectively. Since $\cos\gamma \cdot y_a(c) - \sin\gamma \cdot (fy'_a)(c) = 0 = \cos\gamma \cdot y_b(c) - \sin\gamma \cdot (fy'_b)(c)$, there is a (unique) constant $M \in \mathbb{R}$ such that $y_a(c) = My_b(c)$ and $(fy'_a)(c) = M(fy'_b)(c)$. Thus,

$$y_\#(x) := \begin{cases} y_a(x) & \text{if } x \in (a,c), \\ My_b(x) & \text{if } x \in (c,b) \end{cases} \quad (4.2.1)$$

is a solution to (4.0.1) with $\lambda = \lambda_\#$. Therefore, $\lambda_\#$ is an eigenvalue of the S-L problem consisting of (4.0.1) and (4.0.2) with an eigenfunction $y_\#$. □

Now, we give a geometric characterization of the eigenvalues of the S-L problem consisting of (4.0.1) and (4.0.2).

Theorem 4.2.1 (1) The eigenvalues of the S-L problem consisting of (4.0.1) and (4.0.2) are the λ-coordinates of the intersections of the curves $\lambda = \mu_0(\gamma)$, $\lambda = \mu_{-1}(\gamma)$, $\lambda = \mu_{-2}(\gamma)$, $\cdots$ and $\lambda = \nu_0(\gamma)$, $\lambda = \nu_1(\gamma)$, $\lambda = \nu_2(\gamma)$, $\cdots$ on $[0, \pi)$.

(2) For any $j \geq 0$ and $k \geq 0$, the two curves $\lambda = \mu_{-j}(\gamma)$ and $\lambda = \nu_k(\gamma)$ intersect at most once on $[0, \pi)$.

Proof. (1) This comes directly from Lemma 4.2.1.

(2) The claim is a consequence of the strict increase of μ_{-j} and the strict decrease of ν_k. □

We use Figure 4.2.1 to illustrate the above geometric characterization of the eigenvalues. In the figure, we mark eigenvalues near the corresponding intersections, rather than on the λ-axis. The indices of eigenvalues are determined by their Prüfer angle characterization, see Theorem 4.2.2 and Corollary 4.2.1 below.

Next, as applications of the above characterization, we prove the existence of eigenvalues and give their Prüfer angle characterization.

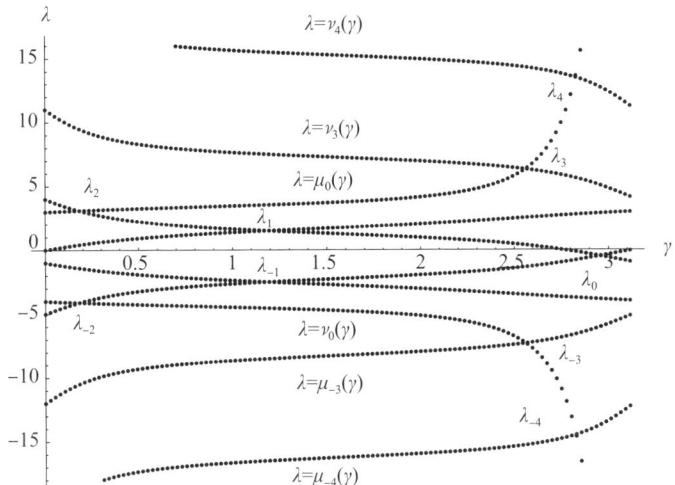

Figure 4.2.1 A geometric characterization of the eigenvalues

Theorem 4.2.2 Consider the S-L problem consisting of (4.0.1) and (4.0.2).

(1) The problem has infinitely many negative eigenvalues approaching $-\infty$ and infinitely many positive eigenvalues accumulating at $+\infty$.

(2) For each $n \in \mathbb{Z}$, there is a unique eigenvalue, to be denoted by λ_n, of the problem such that (4.0.1) with $\lambda = \lambda_n$ has real solutions whose Prüfer angle $\theta(\cdot, \lambda_n)$ satisfies

$$\theta(a, \lambda_n) = \alpha, \qquad \theta(b, \lambda_n) = \beta + n\pi. \tag{4.2.2}$$

(3) The only eigenvalues of the problem are $\{\lambda_n; n \in \mathbb{Z}\}$.

Proof. (1) For each sufficiently large $j \in \mathbb{N}$, the curves $\lambda = \mu_{-j}(\gamma)$ and $\lambda = \nu_0(\gamma)$ intersect in $(0, \pi)$. Thus, the problem has infinitely many negative eigenvalues. From (4.1.1) we know that the eigenvalue coming from the intersection of $\lambda = \mu_{-j}(\gamma)$ and $\lambda = \nu_0(\gamma)$ is $< \mu_{-j+1}(0)$. Thus, by (4.1.2), the negative eigenvalues approach $-\infty$. Similarly, the problem has infinitely many positive eigenvalues, and they accumulate at $+\infty$.

(2) Let $j \subset \mathbb{N}$ be sufficiently large. Then, the intersection of the curves $\lambda = \mu_{-j}(\gamma)$ and $\lambda = \nu_0(\gamma)$ is in $(0, \pi)$, say at $(\gamma, \lambda) = (\gamma_\#, \lambda_\#)$, and hence the Prüfer angle $\theta(\cdot, \lambda_\#)$ of $\lambda_\#$ satisfies

$$\theta(a, \lambda_\#) = \alpha, \quad \theta(c, \lambda_\#) = \gamma_\# + (-j)\pi, \quad \theta(b, \lambda_\#) = \beta + (-j + 0)\pi. \tag{4.2.3}$$

Similarly, for each $k \in \mathbb{N}$ sufficiently large, there is a $\gamma^\# \in (0, \pi)$ such that the problem has an eigenvalue $\lambda^\#$ whose Prüfer angle $\theta(\cdot, \lambda^\#)$ fulfills

$$\theta(a, \lambda^\#) = \alpha, \quad \theta(c, \lambda^\#) = \gamma^\# + 0\pi, \quad \theta(b, \lambda^\#) = \beta + (0 + k)\pi. \tag{4.2.4}$$

Consider the Prüfer angle $\theta(\cdot, \lambda)$ of (4.0.1) determined by $\theta(a, \lambda) = \alpha$. The continuity of $\theta(b, \lambda)$ in λ on $\mathbb{R}$ then implies the existence of λ_n for each $n \in \mathbb{Z}$. Furthermore, the strict increase of $\theta(b, \lambda)$ in λ on $\mathbb{R}$ (see Lemma 4.1.3) yields the uniqueness of each λ_n.

(3) With (2) at hand, this claim is a consequence of Lemma 4.1.1. □

Remark 4.2.1 The above proof of Theorem 4.2.2 can be generalized to proof of Theorem 4.2.2 for the case where f changes the sign only finitely many times. For example, Figure 4.2.2 below shows the situation where f changes the sign exactly twice, from negative to positive first and then from positive to negative. (The point at which f changes from positive to negative is chosen as c, and hence there is a μ_j for each $j \in \mathbb{Z}$ and a ν_{-k} for every integer $k \geqslant 0$. Moreover, the range of the parameter γ is $0 < \gamma \leqslant \pi$, while the range of β is $0 \leqslant \beta < \pi$.) Each problem with a changing sign f can be approximated by problems whose f changes the sign only finitely many times. In this way, one obtains a simple and direct proof of Theorem 4.2.2 for the general case where f changes the sign (i.e., Theorem 1.1 in [102] and Theorem 2.2 in [97]). Here we omit the details.

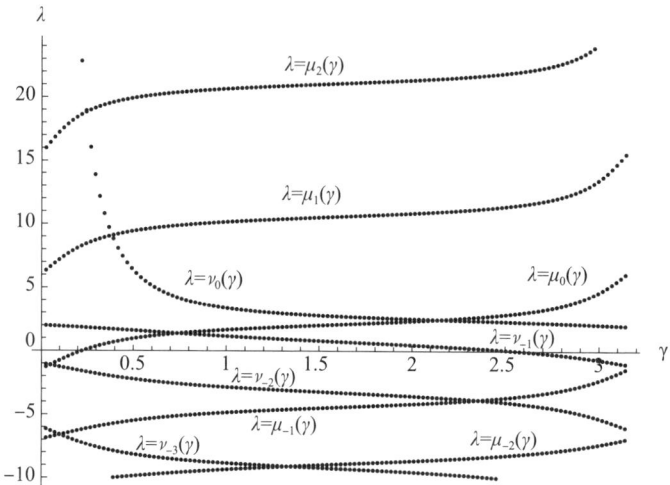

Figure 4.2.2 f changes the sign exactly twice

Finally, we prove a way of determining the index of an eigenvalue, yielded by (4.2.2), from the number of zeros of its eigenfunctions. This result is a slight modification of a special case of Theorem 4.9 in [98].

Theorem 4.2.3 Assume that $n \in \mathbb{Z}$, and y_n is an eigenfunction for the eigenvalue λ_n of the S-L problem consisting of (4.0.1) and (4.0.2). Let n_+ be the number of zeros of y_n in (a, b) having neighborhoods in (a, b) on which $f > 0$ a.e., and $\tilde{n}_-$ the number of zeros of y_n in (a, b) having neighborhoods in (a, b) in which $f < 0$ a.e.. Then,

$$n = n_+ - \tilde{n}_-. \tag{4.2.5}$$

Proof. Assume that λ_n comes from the intersection of the curves $\lambda = \mu_{-j}(\gamma)$ and $\lambda = \nu_k(\gamma)$ above $\gamma = \gamma_n \in [0, \pi)$, where $j \geqslant 0$ and $k \geqslant 0$. Without loss of generality, we assume further that y_n is real-valued. Then, the Prüfer angle $\theta(\cdot, \lambda_n)$ of y_n satisfies that

$$\theta(a, \lambda_n) = \alpha, \quad \theta(c, \lambda_n) = \gamma_n + (-j)\pi, \quad \theta(b, \lambda_n) = \beta + (-j+k)\pi. \tag{4.2.6}$$

Thus, $n = k - j$.

Note that $f < 0$ a.e. on $[a, c]$ and $f > 0$ a.e. on $[c, b]$. The number of zeros of y_n in (a, c) is j, the number in (c, b) is k, $n_+ = k$, $\tilde{n}_- = j$, and hence $n_+ - \tilde{n}_- = k - j = n$. □

The above proof explains the definitions of n_+ and $\tilde{n}_-$ in Theorem 4.2.3. Moreover, the proof implies the following fact.

Corollary 4.2.1 The index of the eigenvalue coming from the intersection of the curves $\lambda = \mu_{-j}(\gamma)$ and $\lambda = \nu_k(\gamma)$ on $[0, \pi)$ is $-j + k$.

Now, we give proof of the corresponding special case of Theorem 4.9 in [98], in which the range of α is chosen as $\alpha \in [0, \pi)$.

Theorem 4.2.4 Assume that $\alpha \in [0, \pi)$ (in this theorem only), $n \in \mathbb{Z}$, λ_n is the eigenvalue of the S-L problem consisting of (4.0.1) and (4.0.2) determined by (4.2.2), and y_n is an eigenfunction for λ_n. Let n_+ be the number of zeros of y_n in (a, b) having neighborhoods in (a, b) on which $f > 0$ a.e., and n_- the number of zeros of y_n on $[a, b]$ having neighborhoods on $[a, b]$ on which $f < 0$ a.e. Then,

$$n = n_+ - n_-. \tag{4.2.7}$$

Proof. Assume that λ_n comes from the intersection of the curves $\lambda = \mu_{-j}(\gamma)$ and $\lambda = \nu_k(\gamma)$ above $\gamma = \gamma_n \in [0, \pi)$, where $j \geqslant 0$ and $k \geqslant 0$. Then, by Corollary 4.2.1, $n = k - j$.

Note that $f < 0$ a.e. on $[a, c]$, $f > 0$ a.e. on $[c, b]$, and $\gamma \in [0, \pi)$. If $\alpha = 0$, then (μ_0 does not exist) a is a zero of y_n, the number of zeros of y_n in (a, c) is $j - 1$, the number in (c, b) is k, $n_+ = k$, $n_- = 1 + (j-1) = j$, and hence $n_+ - n_- = k - j = n$. If $\alpha \in (0, \pi)$, then $n_- = \tilde{n}_-$, and thus (4.2.7) comes from (4.2.5). □

From Theorem 4.2.3 and Theorem 4.2.4 and their above proof we see that for S-L problems whose f changes the sign exactly once and from negative to positive, $\tilde{n}_-$ is more natural than n_-. The price that we have to pay in order to use $\tilde{n}_-$ is choosing $(0, \pi]$ as the range of α.

Remark 4.2.2 We can generalize Theorem 4.2.3 and Theorem 4.2.4 to the case where f changes the sign only finitely many times (i.e., Theorem 4.9 in [98] for such S-L problems). See Remark 4.2.1 for some details.

4.3 Interlacing relations among eigenvalues

We start with partial interlacing relations among eigenvalues.

Theorem 4.3.1 (1) Let $m \in \mathbb{Z}$ satisfy

$$\lambda_m \leqslant \nu_0(0) < \lambda_{m+1}, \tag{4.3.1}$$

then $m \leqslant 0$, and

$$\cdots < \mu_{m-2}(0) < \lambda_{m-2} < \mu_{m-1}(0) < \lambda_{m-1} < \mu_m(0) \leqslant \lambda_m. \tag{4.3.2}$$

(2) Let $n \in \mathbb{Z}$ fulfill

$$\lambda_{n-1} < \mu_0(0) \leqslant \lambda_n, \tag{4.3.3}$$

then $n \geqslant 0$, and

$$\lambda_n \leqslant \nu_n(0) < \lambda_{n+1} < \nu_{n+1}(0) < \lambda_{n+2} < \nu_{n+2}(0) < \cdots. \tag{4.3.4}$$

Proof. (1) Let $i \in \mathbb{Z}$ satisfy $i \leqslant m$. Then $\lambda_i \leqslant \lambda_m$. By (4.3.1) and the strict decrease of each ν_k, λ_i comes from the intersection of the curves $\lambda = \mu_{-j}(\gamma)$ and $\lambda = \nu_0(\gamma)$ on $[0, \pi)$ for some $j \geqslant 0$. From Corollary 4.2.1 we then deduce that $-j = i$. Thus, $m \leqslant 0$, and (4.3.2) follows from the strict increase of μ_k, see Figure 4.3.1 below.

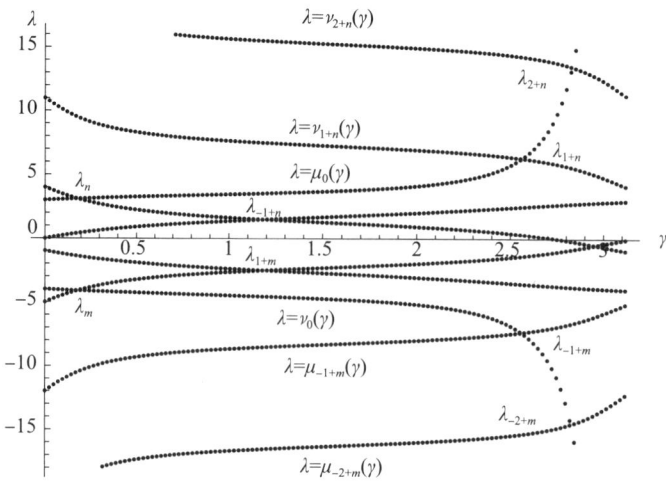

Figure 4.3.1 The strict increase of the μ_k ($m \leqslant 0$)

(2) This part can be shown similarly. □

We use f_+ and f_- to denote the positive and negative parts of f, respectively, such that $f = f_+ - f_-$. As a direct consequence of Theorem 4.3.1 and the asymptotic formulas for the eigenvalues of S-L problems with a definite leading coefficient function, we have the following result. This result is a special case of Theorem 4.1 in [103].

Theorem 4.3.2 The eigenvalues of the S-L problem consisting of (4.0.1) and (4.0.2) satisfy

$$\lambda_n \sim \pm \frac{n^2 \pi^2}{\left[\int_a^b \frac{w(x)}{f_\pm(x)} \, dx\right]^2} \quad \text{as } n \longrightarrow \pm \infty. \tag{4.3.5}$$

Remark 4.3.1 By generalizing the above proof of Theorem 4.3.1 one can get a simple and direct proof of the asymptotic formulas in (4.3.5) for the general case where f changes the sign (i.e., Theorem 4.1 in [103] for such S-L problems).

See Remark 4.2.1 for some details.

Example 4.3.1 Fix an $m \in \mathbb{Z}$ satisfying $m \leqslant -2$, and consider the S-L problem

$$-(fy')' + qy = \lambda y \text{ in } (-\pi, \pi), \tag{4.3.6}$$

$$y(-\pi) = 0 = y(\pi), \tag{4.3.7}$$

where

$$f(x) = \operatorname{sgn}(x) \text{ for } x \in (-\pi, \pi), \quad q(x) = \begin{cases} (m-1)^2 & \text{for } x \in (-\pi, 0), \\ 0 & \text{for } x \in (0, \pi). \end{cases} \tag{4.3.8}$$

Then,

$$\mu_{-j}(0) = (m-1)^2 - (j+1)^2 \text{ for } j \geqslant 0, \quad \nu_k(0) = (k+1)^2 \text{ for } k \geqslant 0. \tag{4.3.9}$$

Note that $-2m + 1 > 4$. This m is equal to the one in Theorem 4.3.1 (1) since $\mu_m(0) = (m-1)^2 - (-m+1)^2 = 0 < 1 = \nu_0(0)$ and $\mu_{m+1}(0) = (m-1)^2 - (-m)^2 = -2m+1 > 1 = \nu_0(0)$. (Hence, $0 < \lambda_m < 1$.) All the numbers $\nu_0(0) = 1$, $\nu_1(0) = 2^2, \cdots$, $\nu_i(0) = (i+1)^2$ are in the interval $(\mu_m(0), \mu_{m+1}(0)) = (0, -2m+1)$, where i is the largest natural number such that $(i+1)^2 < -2m+1$. So the curve $\lambda = \mu_m(\gamma)$ intersects all the curves $\lambda = \nu_0(\gamma)$, $\lambda = \nu_1(\gamma)$, $\cdots$, $\lambda = \nu_i(\gamma)$ on $[0, \pi)$, all the λ-coordinates of the intersections are in $(\mu_m(0), \mu_{m+1}(0))$, and hence all the eigenvalues $\lambda_m, \lambda_{m+1}, \cdots, \lambda_{m+i}$ are in $(\mu_m(0), \mu_{m+1}(0))$. Figure 4.3.2 below shows the case with $m = -3$: λ_{-3} and λ_{-2} are both in $(\mu_{-3}(0), \mu_{-2}(0)) = (0, 7)$. Actually, λ_{-1} is also in the interval. Thus, in general, the partial interlacing relations in Theorem 4.3.1 (1) can not be extended to include even λ_{m+1}.

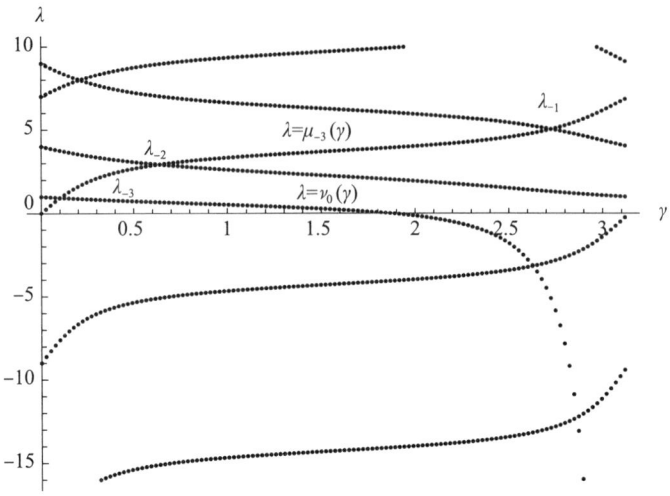

Figure 4.3.2 λ_{-3} and λ_{-2} are both in $(\mu_{-3}(0), \mu_{-2}(0)) = (0, 7)$

Similarly, in general, the partial interlacing relations in Theorem 4.3.1 (2) can not be extended to include even λ_{n-1}.

We can modify the proof of Theorem 4.3.1 to get the following squeezing relations among eigenvalues. Their proof will be omitted.

Theorem 4.3.3 Fix a $\gamma \in (0, \pi)$.
(1) Let $m \in \mathbb{Z}$ satisfy

$$\lambda_m \leqslant \nu_0(\gamma) < \lambda_{m+1}, \tag{4.3.10}$$

then $m \leqslant 0$, and

$$\begin{aligned} \cdots &< \mu_{m-2}(\gamma) < \lambda_{m-2} < \mu_{m-1}(0) \\ &< \mu_{m-1}(\gamma) < \lambda_{m-1} < \mu_m(0) < \mu_m(\gamma) \leqslant \lambda_m. \end{aligned} \tag{4.3.11}$$

(2) Let $n \in \mathbb{Z}$ fulfill

$$\lambda_{n-1} < \mu_0(\gamma) \leqslant \lambda_n, \tag{4.3.12}$$

then $n \geqslant 0$, and

$$\begin{aligned} \lambda_n &\leqslant \nu_n(\gamma) < \nu_n(0) < \lambda_{n+1} < \nu_{n+1}(\gamma) \\ &< \nu_{n+1}(0) < \lambda_{n+2} < \nu_{n+2}(\gamma) < \cdots. \end{aligned} \tag{4.3.13}$$

When $\nu_0(0) < \mu_0(0)$, between $\nu_0(0)$ and $\mu_0(0)$, we need to use all $\mu_{-j}(0)$ and $\nu_k(0)$ there to separate λ_i in this interval, i.e., we have the following full interlacing relations, deduced from Theorem 4.3.1.

Theorem 4.3.4 Let m and n be given by (4.3.1) and (4.3.3).
(1) If $\mu_0(0) \leqslant \nu_0(0)$, then $m = n = 0$, and

$$\cdots < \mu_{-2}(0) < \lambda_{-2} < \mu_{-1}(0) < \lambda_{-1} < \mu_0(0) \leqslant \lambda_0 \leqslant \nu_0(0) < \lambda_1 < \nu_1(0) < \lambda_2 < \nu_2(0) < \lambda_3 < \cdots . \tag{4.3.14}$$

(2) If $\nu_0(0) < \mu_0(0)$, then $m \leqslant -1$, $n \geqslant 1$, and using the non-decreasing list

$$\tau_{m+1} = \nu_0(0) \leqslant \tau_{m+2} \leqslant \cdots \leqslant \tau_{n-1} \leqslant \tau_n = \mu_0(0) \tag{4.3.15}$$

of the eigenvalues $\mu_{m+1}(0), \cdots, \mu_0(0)$ and $\nu_0(0), \cdots, \nu_{n-1}(0)$ together, we have that

$$\cdots < \mu_{m-2}(0) < \lambda_{m-2} < \mu_{m-1}(0) < \lambda_{m-1} < \mu_m(0) \leqslant \lambda_m$$
$$\leqslant \tau_{m+1} \leqslant \lambda_{m+1} \leqslant \cdots \leqslant \tau_{n-1} \leqslant \lambda_{n-1} \leqslant \tau_n \tag{4.3.16}$$
$$\leqslant \lambda_n \leqslant \nu_n(0) < \lambda_{n+1} < \nu_{n+1}(0) < \lambda_{n+2} < \nu_{n+2}(0) < \cdots .$$

Proof. Let $i \in \mathbb{Z}$, and consider λ_i and λ_{i+1}. By Theorem 4.2.1, λ_i comes from the intersection of the curves $\lambda = \mu_{-j}(\gamma)$ and $\lambda = \nu_k(\gamma)$ on $[0, \pi)$ for some $j \geqslant 0$ and $k \geqslant 0$. Then, $\mu_{-j}(0) \leqslant \lambda_i \leqslant \nu_k(0)$. From Corollary 4.2.1 we deduce that $i = -j + k$.

If the intersection of $\lambda = \mu_{-j}(\gamma)$ and $\lambda = \nu_k(\gamma)$ is above $\gamma = 0$, then $\mu_{-j}(0) = \lambda_i = \nu_k(0)$, λ_{i-1} comes from the intersection of $\lambda = \mu_{-j-1}(\gamma)$ and $\lambda = \nu_k(\gamma)$ in $(0, \pi)$, λ_{i+1} corresponds to the intersection of $\lambda = \mu_{-j}(\gamma)$ and $\lambda = \nu_{k+1}(\gamma)$ in $(0, \pi)$, all the $\mu_{-l}(0)$ with $l \neq j$ and all the $\nu_l(0)$ with $l \neq k$ are outside the interval $[\lambda_{i-1}, \lambda_{i+1}]$. Thus, in this case, $\tau_i = \mu_{-j}(0) = \tau_{i+1} = \nu_k(0)$.

For the rest of the proof, we can assume that the intersections yielding λ_i and λ_{i+1} are in $(0, \pi)$. So by Corollary 4.2.1 again, $\mu_{-j}(0) \neq \nu_k(0) \neq \mu_{-j+1}(0)$.

Case 1: $\mu_{-j+1}(0) > \nu_k(0)$. Then, the curve $\lambda = \mu_{-j}(\gamma)$ also meets the curve $\lambda = \nu_{k+1}(\gamma)$ in $(0, \pi)$, yielding λ_{i+1}. Thus, $\nu_k(0)$ is the only member from $\mu_{-l}(0)$ and $\nu_l(0)$ lying on $[\lambda_i, \lambda_{i+1}]$, i.e., $\tau_{i+1} = \nu_k(0)$.

Case 2: $\mu_{-j+1}(0) < \nu_k(0)$. Then, the curve $\lambda = \nu_k(\gamma)$ also meets the curve $\lambda = \mu_{-j+1}(\gamma)$ in $(0, \pi)$, giving λ_{i+1}. Therefore, $\mu_{-j+1}(0)$ is the only number of $\mu_{-l}(0)$ and $\nu_l(0)$ belonging to $[\lambda_i, \lambda_{i+1}]$, i.e., $\tau_{i+1} = \mu_{-j+1}(0)$. □

Putting Theorem 4.3.4 simply: if one lists $\mu_{-j}(0)$ and $\nu_k(0)$ all together in a non-decreasing order, then they interlace λ_i, with all the inequalities before $\mu_m(0)$ and those after $\nu_n(0)$ being always strict.

Chapter 5 Third-order eigenparameter dependent differential operators

In this chapter, we study a third-order symmetric differential equation

$$\ell f := \frac{1}{w}\{-\mathrm{i}[q_0(q_0 f')']' - (p_0 f')' + \mathrm{i}[q_1 f' + (q_1 f)'] + p_1 f\} = \lambda f \text{ on } [a,b], \quad (5.0.1)$$

with boundary conditions

$$L_1 f := (\alpha_1 \lambda + \tilde{\alpha}_1) f(a) - (\alpha_2 \lambda + \tilde{\alpha}_2) f^{[2]}(a) = 0, \quad (5.0.2)$$

$$L_2 f := (\gamma_1 \lambda + \tilde{\gamma}_1) f(b) + (\gamma_2 \lambda + \tilde{\gamma}_2) f^{[2]}(b) = 0, \quad (5.0.3)$$

$$L_3 f := (\sin\theta + \mathrm{i}) f^{[1]}(a) + (\mathrm{i}\sin\theta + 1) f^{[1]}(b) = 0, \quad (5.0.4)$$

where λ is the spectral parameter, $\theta \in (0, \pi]$, q_0, q_1, p_0, p_1, w satisfy the conditions

$$q_0^{-1}, q_0^{-2}, p_0, q_1, p_1, w \in L([a,b], \mathbb{R}), \quad q_0 > 0, \ w > 0. \quad (5.0.5)$$

$\alpha_k, \tilde{\alpha}_k, \gamma_k, \tilde{\gamma}_k (k=1,2)$ are arbitrary real numbers, and satisfy

$$\rho_1 = \tilde{\alpha}_1 \alpha_2 - \alpha_1 \tilde{\alpha}_2 > 0, \quad \rho_2 = \tilde{\gamma}_1 \gamma_2 - \gamma_1 \tilde{\gamma}_2 > 0. \quad (5.0.6)$$

Note that the third-order boundary value problem(BVP) with an eigenparameter contained in the boundary conditions consisting of (5.0.1) ∼ (5.0.4) is considered here.

5.1 Preliminaries

Firstly, we investigate some basic preliminaries.
The quasi-derivatives of f are defined as [127]

$$f^{[0]} = f, \quad f^{[1]} = -\frac{1+\mathrm{i}}{\sqrt{2}} q_0 f', \quad f^{[2]} = \mathrm{i}q_0(q_0 f')' + p_0 f' - \mathrm{i}q_1 f.$$

and $H_w = L_w^2[a,b]$ is a weighted Hilbert space equipped with the inner product $\langle f, g \rangle_w = \int_a^b f\bar{g}w\mathrm{d}x$ consisting of functions f which satisfy

$$\int_a^b |f|^2 w\mathrm{d}x < \infty.$$

The maximal operator $L_{\max}$ is defined as
$$L_{\max} f = \ell f, \ f \in H_w$$
with the domain
$$D_{\max} = \{f \in L_w^2[a,b] \mid f, f^{[1]}, f^{[2]} \in AC[a,b], \ell f \in L_w^2[a,b]\}.$$
Then for arbitrary $f, g \in D_{\max}$, integration by parts yields Lagrange identity
$$\langle L_{\max} f, g \rangle_w - \langle f, L_{\max} g \rangle_w = [f, \overline{g}]_a^b,$$
where
$$[f, \overline{g}]_a^b = [f, \overline{g}](b) - [f, \overline{g}](a),$$
$$[f, \overline{g}](x) = f(x)\overline{g^{[2]}(x)} - f^{[2]}(x)\overline{g(x)} + \mathrm{i} f^{[1]}(x)\overline{g^{[1]}(x)}.$$

Let $\mathcal{H} = L_w^2[a,b] \oplus \mathbb{C}^2$ equipped with inner product
$$\langle \boldsymbol{F}, \boldsymbol{G} \rangle = \int_a^b f \overline{g} w \mathrm{d}x + \frac{1}{\rho_1} f_1 \overline{g_1} + \frac{1}{\rho_2} f_2 \overline{g_2}, \tag{5.1.1}$$

where $\boldsymbol{F} = \begin{pmatrix} f(x) \\ f_1 \\ f_2 \end{pmatrix}, \boldsymbol{G} = \begin{pmatrix} g(x) \\ g_1 \\ g_2 \end{pmatrix} \in \mathcal{H}$. It is easy to verify that $\mathcal{H}$ is a Hilbert space.

Define the operator T as follows
$$\begin{aligned} D(\mathrm{T}) = \{ &F = (f(x), f_1, f_2)^T \in \mathcal{H} \mid L_3 f = 0, \\ &f_1 = \alpha_1 f(a) - \alpha_2 f^{[2]}(a), f_2 = \gamma_1 f(b) + \gamma_2 f^{[2]}(b), f \in D_{\max}\}, \end{aligned} \tag{5.1.2}$$

and
$$\boldsymbol{F} = \begin{pmatrix} f(x) \\ \alpha_1 f(a) - \alpha_2 f^{[2]}(a) \\ \gamma_1 f(b) + \gamma_2 f^{[2]}(b) \end{pmatrix} \in D(\mathrm{T}), \ \mathrm{T}\boldsymbol{F} = \begin{pmatrix} \ell f \\ \widetilde{\alpha}_2 f^{[2]}(a) - \widetilde{\alpha}_1 f(a) \\ -[\widetilde{\gamma}_1 f(b) + \widetilde{\gamma}_2 f^{[2]}(b)] \end{pmatrix}. \tag{5.1.3}$$

By the definition of the operator T, the eigenvalue problem of BVP (5.0.1) $\sim$ (5.0.4) is transferred to the spectra problem of operator T.

Considering the operator T, the following properties hold.

Lemma 5.1.1 T is self-adjoint in $\mathcal{H}$, its eigenvalues are discrete, real-valued, and have no finite point of accumulation. Moreover, the multiplicity of each eigenvalue is at most 3.

Proof. For detailed proof, we refer to [75]. □

Let
$$A_\lambda = \begin{pmatrix} a_1\lambda + \tilde{a}_1 & 0 & -(a_2\lambda + \tilde{a}_2) \\ 0 & 0 & 0 \\ 0 & \sin\theta + i & 0 \end{pmatrix},$$

$$B_\lambda = \begin{pmatrix} 0 & 0 & 0 \\ \gamma_1\lambda + \tilde{\gamma}_1 & 0 & \gamma_2\lambda + \tilde{\gamma}_2 \\ 0 & i\sin\theta + 1 & 0 \end{pmatrix}.$$

Let $\vartheta_1(x,\lambda), \vartheta_2(x,\lambda), \vartheta_3(x,\lambda)$ be the system of linearly independent fundamental solutions to equation (5.0.1) and satisfy the initial condition

$$\Theta(x,\lambda) = \begin{pmatrix} \vartheta_1^{[0]}(x,\lambda) & \vartheta_2^{[0]}(x,\lambda) & \vartheta_3^{[0]}(x,\lambda) \\ \vartheta_1^{[1]}(x,\lambda) & \vartheta_2^{[1]}(x,\lambda) & \vartheta_3^{[1]}(x,\lambda) \\ \vartheta_1^{[2]}(x,\lambda) & \vartheta_2^{[2]}(x,\lambda) & \vartheta_3^{[2]}(x,\lambda) \end{pmatrix}, \quad \Theta(a,\lambda) = \begin{pmatrix} 1 & 0 & 0 \\ 0 & 1 & 0 \\ 0 & 0 & 1 \end{pmatrix}. \quad (5.1.4)$$

Lemma 5.1.2 $\lambda \in \mathbb{C}$ is an eigenvalue of BVP (5.0.1) $\sim$ (5.0.4) (also of the operator T) if and only if
$$\Delta(\lambda) = \det[A_\lambda + B_\lambda \Theta(b,\lambda)] = 0.$$

Proof. The proof is routine by substituting the general solution consisting of the independent solutions $\vartheta_1(x,\lambda), \vartheta_2(x,\lambda), \vartheta_3(x,\lambda)$ into boundary conditions (5.0.2) $\sim$ (5.0.4), then we can get a system of homogeneous linear equations, the using ordinary differential equation theory, we can get the conclusion. □

5.2 The Banach space

For the purpose of investigating the continuity of eigenvalues, we introduce the following Banach space.

Consider the Banach space
$$\mathcal{W} = L[a,b] \oplus L[a,b] \oplus L[a,b] \oplus \mathbb{R}^9$$

with the norm
$$\|\omega\| = \int_a^b (|p_0| + |p_1| + |w|)\mathrm{d}x + |\Delta_1| + |\Delta_2| + |\theta|$$

for any $W = (p_0, p_1, w, \Delta_1, \Delta_2, \theta) \in \mathcal{W}$, where

$$|\Delta_1| = |\alpha_1| + |\alpha_2| + |\tilde{\alpha}_1| + |\tilde{\alpha}_2|, |\Delta_2| = |\gamma_1| + |\gamma_2| + |\tilde{\gamma}_1| + |\tilde{\gamma}_2|.$$

Let
$$\Omega = \{\omega \in \mathcal{W} : (5.0.5), (5.0.6) \text{ hold, and } \theta \in (0, \pi]\}.$$

Lemma 5.2.1 The solution $f = f(x, c_0, d_1, d_2, d_3, p_0, p_1, w)$ to the equation (5.0.1) satisfying the conditions
$$f(c_0, \lambda) = d_1, \ f^{[1]}(c_0, \lambda) = d_2, \ f^{[2]}(c_0, \lambda) = d_3, \ c_0 \in [a, b]$$
is continuous with respect to all its variables.

Proof. By transferring the equation (5.0.1) to a first-order system, the conclusion holds based on the existence and uniqueness of solutions and Theorem 2.7 in [128]. □

Theorem 5.2.1 Let $\omega^* = (p_0^*, p_1^*, w^*, \Delta_1^*, \Delta_2^*, \theta^*) \in \Omega$, and $\lambda = \lambda(\omega)$ be an eigenvalue of BVP (5.0.1) $\sim$ (5.0.4). Then for any $\varepsilon > 0$, there exists a $\delta > 0$ so that if $\omega \in \Omega$ satisfies
$$\|\omega - \omega^*\| = \int_a^b (|p_0 - p_0^*| + |p_1 - p_1^*| + |w - w^*|) \mathrm{d}x +$$
$$|\alpha_1 - \alpha_1^*| + |\alpha_2 - \alpha_2^*| + |\tilde{\alpha}_1 - \tilde{\alpha}_1^*| + |\tilde{\alpha}_2 - \tilde{\alpha}_2^*| +$$
$$|\gamma_1 - \gamma_1^*| + |\gamma_2 - \gamma_2^*| + |\tilde{\gamma}_1 - \tilde{\gamma}_1^*| + |\tilde{\gamma}_2 - \tilde{\gamma}_2^*| + |\theta - \theta^*|$$
$$< \delta,$$

then
$$|\lambda(\omega) - \lambda(\omega^*)| < \varepsilon.$$

Proof. By Lemma 5.1.2, one can see that λ is an eigenvalue of BVP (5.0.1) $\sim$ (5.0.4) if and only if $\Delta(\lambda) = 0$. By the continuous dependence of solutions on the problem [128], $\Delta(\lambda)$ is an entire function with respect to λ and is continuous at $\omega \in \Omega$. We can also verify that $\Delta(\lambda)$ is not a constant. According to the continuity theorem for analytic functions with isolated zeros, the conclusion holds. □

Definition 5.2.1 We call $\boldsymbol{F} = (f(x), f_1, f_2)^T \in H$ is an normalized eigenvector of the operator T, where $f_1 = \alpha_1 f(a) - \alpha_2 f^{[2]}(a)$, $f_2 = \gamma_1 f(b) + \gamma_2 f^{[2]}(b)$, if the eigenvector $\boldsymbol{F} = (f(x), f_1, f_2)^T$ of T satisfies
$$\|\boldsymbol{F}\|^2 = \langle (f(x), f_1, f_2)^T, (f(x), f_1, f_2)^T \rangle$$
$$= \int_a^b |f(x)|^2 w \mathrm{d}x + \frac{1}{\rho_1}|f_1|^2 + \frac{1}{\rho_2}|f_2|^2$$
$$= 1.$$

The continuity of eigenvectors of the operator T is stated as follows.

Theorem 5.2.2 Let $\lambda(\omega)(\omega \in \Omega)$ be an eigenvalue with multiplicity n $(n = 1, 2, 3)$ for all ω in some neighborhoods of ω^* in Ω. Let $\boldsymbol{F}_k(x, \omega^*) = \big(f_k(x, \omega^*), f_{k1}(\omega^*), f_{k2}(\omega^*)\big)^T$ be the normalized eigenvectors of $\lambda(\omega^*)$. Then, there exist n linearly independent normalized eigenvectors $\boldsymbol{F}_k(x, \omega) = \big(f_k(x, \omega), f_{k1}(\omega), f_{k2}(\omega)\big)^T$ of $\lambda(\omega)$ so that

$$f_k(x,\omega) \to f_k(x,\omega^*),\ f_k^{[1]}(x,\omega) \to f_k^{[1]}(x,\omega^*),\ f_k^{[2]}(x,\omega) \to f_k^{[2]}(x,\omega^*),$$

$$f_{k1}(\omega) \to f_{k1}(\omega^*),\ f_{k2}(\omega) \to f_{k2}(\omega^*),\ k = 1, \cdots, n, \text{ as } \omega \to \omega^* \text{ in } \Omega$$

both uniformly on $[a, b]$.

Proof. It can be proved with similar methods in [52], with the aid of Lemma 5.2.1, Theorem 5.2.1 as well as Theorem 3.2 in [46]. □

5.3 Derivative formulas of eigenvalues

We give the differentiability of eigenvalues concerning some data, especially the boundary condition parameter matrix in this section.

Definition 5.3.1[129] Let $\mathbf{X}$, $\mathbf{Y}$ be Banach spaces, we call the map $\Gamma : \mathbf{X} \to \mathbf{Y}$ is Fréchet differentiable at $x \in \mathbf{X}$ provided that a bounded linear operator $\mathrm{d}\Gamma_x : \mathbf{X} \to \mathbf{Y}$ exists, such that for $\tau \in \mathbf{X}$,

$$|\Gamma(x + \tau) - \Gamma(x) - \mathrm{d}\Gamma_x(\tau)| = o(\tau),\ \tau \to 0.$$

Theorem 5.3.1 Let $\omega = (p_0, p_1, w, \Delta_1, \Delta_2, \theta) \in \Omega$, $\lambda = \lambda(\omega)$ be an eigenvalue of BVP (5.0.1) $\sim$ (5.0.4), $\boldsymbol{F}(\omega) = \big(f(x,\omega), f_1(\omega), f_2(\omega)\big)^T \in \mathcal{H}$ be the corresponding eigenvector. $\boldsymbol{E}$ is the identity matrix, $\boldsymbol{R}$ is 2×2 real-valued matrix. Assume that for all fixed elements of ω except 1, the geometric multiplicity of $\lambda(\omega)$ is invariant in some neighborhoods $\mathcal{M} \subset \Omega$. Then the following differential expressions hold.

(1) Let all the elements of ω be fixed except θ and $\lambda = \lambda(\theta)$. Then λ is differentiable and satisfies

$$\lambda'(\theta) = \frac{2\cos\theta}{1 + \sin^2\theta}|f^{[1]}(a)|^2 = \frac{2\cos\theta}{1 + \sin^2\theta}|f^{[1]}(b)|^2.$$

(2) Let all the elements of ω be fixed except α_1 and $\lambda = \lambda(\alpha_1)$. Then λ is differentiable and satisfies

$$\lambda'(\alpha_1) = \frac{\lambda}{\alpha_2 \lambda + \widetilde{\alpha}_2} |f(a)|^2,$$

where $\alpha_2 \lambda + \widetilde{\alpha}_2 \neq 0$.

(3) Let all the elements of ω be fixed except $\widetilde{\alpha}_1$ and $\lambda = \lambda(\widetilde{\alpha}_1)$. Then λ is differentiable and satisfies

$$\lambda'(\widetilde{\alpha}_1) = \frac{1}{\alpha_2 \lambda + \widetilde{\alpha}_2} |f(a)|^2,$$

where $\alpha_2 \lambda + \widetilde{\alpha}_2 \neq 0$.

(4) Let all the elements of ω be fixed except α_2 and $\lambda = \lambda(\alpha_2)$. Then λ is differentiable and satisfies

$$\lambda'(\alpha_2) = -\frac{\lambda}{\alpha_1 \lambda + \widetilde{\alpha}_1} |f^{[2]}(a)|^2,$$

where $\alpha_1 \lambda + \widetilde{\alpha}_1 \neq 0$.

(5) Let all the elements of ω be fixed except $\widetilde{\alpha}_2$ and $\lambda = \lambda(\widetilde{\alpha}_2)$. Then λ is differentiable and satisfies

$$\lambda'(\widetilde{\alpha}_2) = -\frac{1}{\alpha_1 \lambda + \widetilde{\alpha}_1} |f^{[2]}(a)|^2,$$

where $\alpha_1 \lambda + \widetilde{\alpha}_1 \neq 0$.

(6) Let all the elements of ω be fixed except γ_1 and $\lambda = \lambda(\gamma_1)$. Then λ is differentiable and satisfies

$$\lambda'(\gamma_1) = \frac{\lambda}{\gamma_2 \lambda + \widetilde{\gamma}_2} |f(b)|^2,$$

where $\gamma_2 \lambda + \widetilde{\gamma}_2 \neq 0$.

(7) Let all the elements of ω be fixed except $\widetilde{\gamma}_1$ and $\lambda = \lambda(\widetilde{\gamma}_1)$. Then λ is differentiable and satisfies

$$\lambda'(\widetilde{\gamma}_1) = \frac{1}{\gamma_2 \lambda + \widetilde{\gamma}_2} |f(b)|^2,$$

where $\gamma_2 \lambda + \widetilde{\gamma}_2 \neq 0$.

(8) Let all the elements of ω be fixed except γ_2 and $\lambda = \lambda(\gamma_2)$. Then λ is differentiable and satisfies

$$\lambda'(\gamma_2) = -\frac{\lambda}{\gamma_1 \lambda + \widetilde{\gamma}_1} |f^{[2]}(b)|^2,$$

where $\gamma_1 \lambda + \widetilde{\gamma}_1 \neq 0$.

(9) Let all the elements of ω be fixed except $\widetilde{\gamma}_2$ and $\lambda = \lambda(\widetilde{\gamma}_2)$. Then λ is differentiable and satisfies

$$\lambda'(\widetilde{\gamma}_2) = -\frac{1}{\gamma_1 \lambda + \widetilde{\gamma}_1} |f^{[2]}(b)|^2,$$

where $\gamma_1 \lambda + \widetilde{\gamma}_1 \neq 0$.

(10) Let all the elements of ω be fixed except the boundary condition parameter matrix $K = \begin{pmatrix} \alpha_1 & \tilde{\alpha}_1 \\ \alpha_2 & \tilde{\alpha}_2 \end{pmatrix}$ and $\lambda = \lambda(K)$. Then for all R satisfying $\det[K + R] = -\rho_1$, the Fréchet derivative of λ with respect to K is formulated as

$$d\lambda_K(R) = (-f(a), f^{[2]}(a))[E - (K + R)K^{-1}](\overline{f^{[2]}(a)}, \overline{f(a)})^T.$$

(11) Let all the elements of ω be fixed except the boundary condition parameter matrix $\widetilde{K} = \begin{pmatrix} \gamma_1 & \tilde{\gamma}_1 \\ \gamma_2 & \tilde{\gamma}_2 \end{pmatrix}$ and $\lambda = \lambda(\widetilde{K})$. Then for all R satisfying $\det[\widetilde{K} + R] = -\rho_2$, the Fréchet derivative of λ with respect to $\widetilde{K}$ is formulated as

$$d\lambda_{\widetilde{K}}(R) = (f(b), f^{[2]}(b))[E - (\widetilde{K} + R)\widetilde{K}^{-1}](\overline{f^{[2]}(b)}, -\overline{f(b)})^T.$$

(12) Let all the elements of ω be fixed except p_0 and $\lambda = \lambda(p_0)$. Then the Fréchet derivative of λ with respect to p_0 is formulated as

$$d\lambda_{p_0}(\tau) = \int_a^b \tau |f^{[1]}|^2 dx, \quad \tau \in L[a, b].$$

(13) Let all the elements of ω be fixed except p_1 and $\lambda = \lambda(p_1)$. Then the Fréchet derivative of λ with respect to p_1 is formulated as

$$d\lambda_{p_1}(\tau) = \int_a^b \tau |f|^2 dx, \quad \tau \in L[a, b].$$

(14) Let all the elements of ω be fixed except w and $\lambda = \lambda(w)$. Then the Fréchet derivative of λ with respect to w is formulated as

$$d\lambda_w(\tau) = \lambda_w \int_a^b \tau |f|^2 dx, \quad \tau \in L[a, b].$$

Proof. Let all the elements of $\omega \in \Omega$ be fixed except 1. For sufficiently small $\varepsilon > 0$, when $\|\omega - \omega^*\| < \varepsilon$, let $\lambda(\omega)$ be an eigenvalue of BVP (5.0.1) $\sim$ (5.0.4) satisfying Theorem 5.2.1. For all the above six cases, $\lambda(\omega)$ is replaced with $\lambda(\theta+\Delta\theta)$, $\lambda(\alpha_1+\Delta\alpha_1)$, $\lambda(\tilde{\alpha}_1 + \Delta\tilde{\alpha}_1)$, $\lambda(\alpha_2 + \Delta\alpha_2)$, $\lambda(\tilde{\alpha}_2 + \Delta\tilde{\alpha}_2)$, $\lambda(\gamma_1 + \Delta\gamma_1)$, $\lambda(\tilde{\gamma}_1 + \Delta\tilde{\gamma}_1)$, $\lambda(\gamma_2 + \Delta\gamma_2)$, $\lambda(\tilde{\gamma}_2 + \Delta\tilde{\gamma}_2)$, $\lambda(K + R)$, $\lambda(\widetilde{K} + R)$, $\lambda(p_0 + \eta)$, $\lambda(p_1 + \eta)$, $\lambda(w + \eta)$, respectively.

(1) Let all the elements of ω be fixed except θ. We adopt the following notations:

$$\mathcal{B}_1(f) = \alpha_1 f(a) - \alpha_2 f^{[2]}(a), \quad \mathcal{B}_2(f) = \gamma_1 f(b) + \gamma_2 f^{[2]}(b),$$

$$\mathcal{D}_1(f) = \tilde{\alpha}_2 f^{[2]}(a) - \tilde{\alpha}_1 f(a), \quad \mathcal{D}_2(f) = -[\tilde{\gamma}_1 f(b) + \tilde{\gamma}_2 f^{[2]}(b)].$$

Let

$$F(x,\theta) = \big(f(x,\theta), f_1(\theta), f_2(\theta)\big)^T,$$
$$G(x,\theta) = \big(g(x,\theta), g_1(\theta), g_2(\theta)\big)^T$$

be the normalized eigenvectors corresponding to $\lambda(\theta)$ and $\lambda(\theta+\Delta\theta)$ respectively. By the self-adjointness of the operator T and the boundary condition (5.0.4), we have

$$[\lambda(\theta+\Delta\theta) - \lambda(\theta)]\langle G, F\rangle$$
$$=\langle \lambda(\theta+\Delta\theta)G, F\rangle - \langle G, \lambda(\theta)F\rangle$$
$$=[g,\overline{f}]_a^b + \frac{1}{\rho_1}[\mathcal{D}_1(g)\overline{\mathcal{B}_1(f)} - \mathcal{B}_1(g)\overline{\mathcal{D}_1(f)}] + \frac{1}{\rho_2}[\mathcal{D}_2(g)\overline{\mathcal{B}_2(f)} - \mathcal{B}_2(g)\overline{\mathcal{D}_2(f)}] \quad (5.3.1)$$
$$= ig^{[1]}(b)\overline{f^{[1]}(b)} - ig^{[1]}(a)\overline{f^{[1]}(a)}$$
$$= i\left[\frac{i+\sin(\theta+\Delta\theta)}{1+i\sin(\theta+\Delta\theta)}\frac{\sin\theta-i}{1-i\sin\theta} - 1\right]g^{[1]}(a)\overline{f^{[1]}(a)}.$$

Dividing both sides of $\Delta\theta$, and taking the limit as $\Delta\theta \to 0$, we have

$$\lambda'(\theta) = \frac{2\cos\theta}{1+\sin^2\theta}|f^{[1]}(a)|^2 \quad (5.3.2)$$

by Theorem 5.2.2. Using the boundary condition (5.0.4), we have $|f^{[1]}(a)|^2 = |f^{[1]}(b)|^2$. Hence, (1) holds.

(2) Let all the elements of ω be fixed except α_1, and

$$F(x,\alpha_1) = (f(x,\alpha_1), f_1(\alpha_1), f_2(\alpha_1))^T,$$
$$G(x,\alpha_1) = (g(x,\alpha_1), g_1(\alpha_1), g_2(\alpha_1))^T$$

be the normalized eigenvectors corresponding to $\lambda(\alpha_1)$ and $\lambda(\alpha_1+\Delta\alpha_1)$, respectively. By the self-adjointness of the operator T and the boundary condition (5.0.2), (5.0.4), we have

$$[\lambda(\alpha_1+\Delta\alpha_1) - \lambda(\alpha_1)]\langle G, F\rangle$$
$$=\langle \lambda(\alpha_1+\Delta\alpha_1)G, F\rangle - \langle G, \lambda(\alpha_1)F\rangle$$
$$=[y,\overline{f}]_a^b + \frac{1}{\rho_1}[\mathcal{D}_1(g)\overline{\mathcal{B}_1(f)} - \mathcal{B}_1(g)\overline{\mathcal{D}_1(f)}] + \frac{1}{\rho_2}[\mathcal{D}_2(g)\overline{\mathcal{B}_2(f)} - \mathcal{B}_2(g)\overline{\mathcal{D}_2(f)}]$$
$$= g^{[2]}(a)\overline{f(a)} - g(a)\overline{f^{[2]}(a)} + \frac{1}{\rho_1}[\widetilde{\alpha}_2 g^{[2]}(a) - \widetilde{\alpha}_1 g(a)][\alpha_1\overline{f(a)} - \alpha_2\overline{f^{[2]}(a)}] -$$
$$\frac{1}{\rho_1}[(\alpha_1+\Delta\alpha_1)g(a) - \alpha_2 g^{[2]}(a)][\widetilde{\alpha}_2\overline{f^{[2]}(a)} - \widetilde{\alpha}_1\overline{f(a)}]$$
$$= \frac{\Delta\alpha_1}{\rho_1}(\widetilde{\alpha}_1 g(a)\overline{f(a)} - \widetilde{\alpha}_2 g(a)\overline{f^{[2]}(a)})$$

$$=\frac{\Delta\alpha_1}{\rho_1}(\tilde{\alpha}_1 - \tilde{\alpha}_2\frac{a_1\lambda + \tilde{a}_1}{a_2\lambda + \tilde{a}_2})g(a)\overline{f(a)}$$

$$=\frac{\lambda\Delta\alpha_1}{a_2\lambda + \tilde{a}_2}g(a)\overline{f(a)}. \tag{5.3.3}$$

Dividing both sides of $\Delta\alpha_1$, and taking the limit as $\Delta\alpha_1 \to 0$, we have

$$\lambda'(\alpha_1) = \frac{\lambda}{a_2\lambda + \tilde{a}_2}|f(a)|^2 \tag{5.3.4}$$

by Theorem 5.2.2.

Using the same method, one can prove that (3), (4) and (5) are also true.

(6) Let all the elements of ω be fixed except γ_1, and

$$F(x, \gamma_1) = (f(x, \gamma_1), f_1(\gamma_1), f_2(\gamma_1))^T,$$

$$G(x, \gamma_1) = (g(x, \gamma_1), g_1(\gamma_1), g_2(\gamma_1))^T$$

be the normalized eigenvectors corresponding to $\lambda(\gamma_1)$ and $\lambda(\gamma_1 + \Delta\gamma_1)$, respectively. Then by the self-adjointness of the operator T and the boundary conditions (5.0.3), (5.0.4), we have

$$[\lambda(\gamma_1 + \Delta\gamma_1) - \lambda(\gamma_1)]\langle G, F\rangle$$

$$=\langle \lambda(\gamma_1 + \Delta\gamma_1)G, F\rangle - \langle G, \lambda(\gamma_1)F\rangle$$

$$=[g, \overline{f}]_a^b + \frac{1}{\rho_1}[\mathcal{D}_1(g)\overline{\mathcal{B}_1(f)} - \mathcal{B}_1(g)\overline{\mathcal{D}_1(f)}] + \frac{1}{\rho_2}[\mathcal{D}_2(g)\overline{\mathcal{B}_2(f)} - \mathcal{B}_2(g)\overline{\mathcal{D}_2(f)}]$$

$$=g(b)\overline{f^{[2]}(b)} - g^{[2]}(b)\overline{f(b)} - \frac{1}{\rho_2}[\tilde{\gamma}_1 g(b) + \tilde{\gamma}_2 g^{[2]}(b)][\gamma_1\overline{f(b)} + \gamma_2\overline{f^{[2]}(b)}]+$$

$$\frac{1}{\rho_2}[(\gamma_1 + \Delta\gamma_1)g(b) + \gamma_2 g^{[2]}(b)][\tilde{\gamma}_1\overline{f(b)} + \tilde{\gamma}_2\overline{f^{[2]}(b)}] \tag{5.3.5}$$

$$=\frac{\Delta\gamma_1}{\rho_2}(\tilde{\gamma}_1 g(b)\overline{f(b)} + \tilde{\gamma}_2 g(b)\overline{f^{[2]}(b)})$$

$$=\frac{\Delta\gamma_1}{\rho_2}(\tilde{\gamma}_1 - \tilde{\gamma}_2\frac{\gamma_1\lambda + \tilde{\gamma}_1}{\gamma_2\lambda + \tilde{\gamma}_2})g(a)\overline{f(a)}$$

$$=\frac{\lambda\Delta\gamma_1}{\gamma_2\lambda + \tilde{\gamma}_2}g(b)\overline{f(b)}.$$

Dividing both sides of $\Delta\gamma_1$, and taking the limit as $\Delta\gamma_1 \to 0$, we have

$$\lambda'(\gamma_1) = \frac{\lambda}{\gamma_2\lambda + \tilde{\gamma}_2}|f(b)|^2 \tag{5.3.6}$$

by Theorem 5.2.2.

Using the similar method, one can prove that (7), (8) and (9) are also true.

(10) Let all the elements of ω be fixed except K. Let $K + R = \begin{pmatrix} \alpha_{1R} & \tilde{\alpha}_{1R} \\ \alpha_{2R} & \tilde{\alpha}_{2R} \end{pmatrix}$, with $\det(K + R) = -\rho_1$, and

$$F(x, K) = (f(x, K), f_1(K), f_2(K))^T,$$
$$G(x, K) = (g(x, K), g_1(K), g_2(K))^T$$

be the normalized eigenvectors corresponding to $\lambda(K)$ and $\lambda(K+R)$, respectively. Then

$$\lambda(K)[\alpha_1 f(a) - \alpha_2 f^{[2]}(a)] = \tilde{\alpha}_2 f^{[2]}(a) - \tilde{\alpha}_1 f(a),$$
$$\lambda(K + R)[\alpha_{1R}\overline{g(a)} - \alpha_{2R}\overline{g^{[2]}(a)}] = \tilde{\alpha}_{2R}\overline{g^{[2]}(a)} - \tilde{\alpha}_{1R}\overline{g(a)}$$

by the boundary condition (5.0.2). Using the boundary conditions (5.0.2), (5.0.3) and (5.0.4), simple calculation yields

$$[\lambda(K + R) - \lambda(K)]\langle G, F \rangle$$
$$= \langle \lambda(K + R)G, F \rangle - \langle G, \lambda(K)F \rangle$$
$$= [g, \overline{f}]_a^b + \frac{1}{\rho_1}\mathcal{D}_1(g)\overline{\mathcal{B}_1(f)} - \frac{1}{\rho_1}\mathcal{B}_1(g)\overline{\mathcal{D}_1(f)} + \frac{1}{\rho_2}\mathcal{D}_2(g)\overline{\mathcal{B}_2(f)} - \frac{1}{\rho_2}\mathcal{B}_2(g)\overline{\mathcal{D}_2(f)}$$
$$= g^{[2]}(a)\overline{f(a)} - g(a)\overline{f^{[2]}(a)} + \frac{1}{\rho_1}\mathcal{D}_1(g)\overline{\mathcal{B}_1(f)} - \frac{1}{\rho_1}\mathcal{B}_1(g)\overline{\mathcal{D}_1(f)}$$
$$= g^{[2]}(a)\overline{f(a)} - g(a)\overline{f^{[2]}(a)} + \frac{1}{\rho_1}[\tilde{\alpha}_{2R}g^{[2]}(a) - \tilde{\alpha}_{1R}g(a)][\alpha_1\overline{f(a)} - \alpha_2\overline{f^{[2]}(a)}] -$$
$$\frac{1}{\rho_1}[\alpha_{1R}g(a) - \alpha_{2R}g^{[2]}(a)][\tilde{\alpha}_2\overline{f^{[2]}(a)} - \tilde{\alpha}_1\overline{f(a)}]$$
$$= (-g(a), g^{[2]}(a))(\overline{f^{[2]}(a)}, \overline{f(a)})^T +$$
$$\frac{1}{\rho_1}(-g(a), g^{[2]}(a))(\tilde{\alpha}_{1R}, \tilde{\alpha}_{2R})^T(-\alpha_2, \alpha_1)(\overline{f^{[2]}(a)}, \overline{f(a)})^T -$$
$$\frac{1}{\rho_1}(-g(a), g^{[2]}(a))(-\alpha_{1R}, -\alpha_{2R})^T(\tilde{\alpha}_2, -\tilde{\alpha}_1)(\overline{f^{[2]}(a)}, \overline{f(a)})^T$$
$$= (-g(a), g^{[2]}(a))[E + \frac{1}{\rho_1}(\tilde{\alpha}_{1R}, \tilde{\alpha}_{2R})^T(-\alpha_2, \alpha_1) -$$
$$\frac{1}{\rho_1}(-\alpha_{1R}, -\alpha_{2R})^T(\tilde{\alpha}_2, -\tilde{\alpha}_1)](\overline{f^{[2]}(a)}, \overline{f(a)})^T$$
$$= (-g(a), g^{[2]}(a))\left[E + \frac{1}{\rho_1}\begin{pmatrix} \alpha_{1R}\tilde{\alpha}_2 - \tilde{\alpha}_{1R}\alpha_2 & \tilde{\alpha}_{1R}\alpha_1 - \alpha_{1R}\tilde{\alpha}_1 \\ \alpha_{2R}\tilde{\alpha}_2 - \tilde{\alpha}_{2R}\alpha_2 & \tilde{\alpha}_{2R}\alpha_1 - \alpha_{2R}\tilde{\alpha}_1 \end{pmatrix}\right](\overline{f^{[2]}(a)}, \overline{f(a)})^T$$
$$= (-g(a), g^{[2]}(a))[E - (K + R)K^{-1}](\overline{f^{[2]}(a)}, \overline{f(a)})^T.$$

Let $R \to 0$, we get that

$$[\lambda(K+R) - \lambda(K)](1+o(1)) = (-f(a), f^{[2]}(a))[E - (K+R)K^{-1}](\overline{f^{[2]}(a)}, \overline{f(a)})^T,$$

that is

$$\lambda(K+R) - \lambda(K) = (-f(a), f^{[2]}(a))[E - (K+R)K^{-1}](\overline{f^{[2]}(a)}, \overline{f(a)})^T + o(R).$$

Hence (10) holds.

Proof of (11) is similar to this.

(12) Let all the elements of ω be fixed except p_0, and

$$F(x, p_0) = (f(x, p_0), f_1(p_0), f_2(p_0))^T,$$

$$G(x, p_0) = (g(x, p_0), g_1(p_0), g_2(p_0))^T$$

be the normalized eigenvectors corresponding to $\lambda(p_0)$ and $\lambda(p_0 + \tau)$. Then we have

$$[\lambda(p_0 + \tau) - \lambda(p_0)]\langle G, F \rangle$$

$$= [\lambda(p_0 + \tau) - \lambda(p_0)](\int_a^b g\overline{f}w\mathrm{d}x + \frac{1}{\rho_1}g_1\overline{f_1} + \frac{1}{\rho_2}g_2\overline{f_2})$$

$$= \int_a^b \ell(g)\overline{f}w\mathrm{d}x - \int_a^b g\overline{\ell(f)}w\mathrm{d}x + [\lambda(p_0+\tau) - \lambda(p_0)](\frac{1}{\rho_1}g_1\overline{f_1} + \frac{1}{\rho_2}g_2\overline{f_2})$$

$$= \int_a^b [(-g^{[2]})' + iq_1 g' + p_1 g]\overline{f}\mathrm{d}x - \int_a^b g\overline{[(-f^{[2]})' + iq_1 f' + p_1 f]}\mathrm{d}x +$$

$$\frac{1}{\rho_1}\mathcal{D}_1(g)\overline{\mathcal{B}_1(f)} - \frac{1}{\rho_1}\mathcal{B}_1(g)\overline{\mathcal{D}_1(f)} + \frac{1}{\rho_2}\mathcal{D}_2(g)\overline{\mathcal{B}_2(f)} - \frac{1}{\rho_2}\mathcal{B}_2(g)\overline{\mathcal{D}_2(f)}$$

$$= [g\overline{f^{[2]}} - g^{[2]}\overline{f}]\,|_a^b + \int_a^b [iq_0(q_0 g')' + (p_0+\tau)g' - iq_1 g]\overline{f}'\mathrm{d}x -$$

$$\int_a^b g'\overline{[-iq_0(q_0 f')' + p_0 f' + iq_1 f]}\mathrm{d}x + i\int_a^b [q_1 g'\overline{f} + q_1 g\overline{f}']\mathrm{d}x +$$

$$\frac{1}{\rho_1}\mathcal{D}_1(g)\overline{\mathcal{B}_1(f)} - \frac{1}{\rho_1}\mathcal{B}_1(g)\overline{\mathcal{D}_1(f)} + \frac{1}{\rho_2}\mathcal{D}_2(g)\overline{\mathcal{B}_2(f)} - \frac{1}{\rho_2}\mathcal{B}_2(g)\overline{\mathcal{D}_2(f)}$$

$$= \int_a^b (p_0+\tau)g'\overline{f}'\mathrm{d}x - \int_a^b p_0 g'\overline{f}'\mathrm{d}x$$

$$= \int_a^b \tau g'\overline{f}'\mathrm{d}x.$$

It follows from the above results that (12) holds.

Using the similar method, one can prove that (13) and (14) are also true. □

Chapter 6 Application of Sturm-Liouville problems

6.1 Construction and stability of Riesz bases

As is well known that Riesz basis is not only a base but also a special frame. The research of frame and Riesz basis play an important role in theoretical research of wavelet analysis [130]. Because of the redundancy of frame and Riesz basis, they have been extensively applied in signal denoising, feature extraction, robust signal processing and so on. Therefore, the construction of Riesz basis has attracted much attention from researchers.

In this section, first, we construct two groups of Riesz bases $\{1\} \cup \{\cos(2nx)\} \cup \{\sin(2nx)\}$ and $\{\sin((2n-1)x)\} \cup \{\cos((2n-1)x)\}$, and study their stability. Then, we consider the problem of finding a new sequence associated with eigenfunctions of the S-L problem

$$\begin{cases} -y'' + qy = \lambda y \text{ on } [0, \pi], \\ y(0) = y(\pi) = 0, \end{cases} \quad (6.1.1)$$

such that it forms a Riesz basis.

See [109] for the contents of this section.

6.1.1 Basic concepts and lemmas

Let us first recall some basic concepts. Let $\{f_n\}, n \in \mathbb{N}$, be a sequence in a Hilbert space H, where $\mathbb{N}$ is the set of positive integers. The sequence is called complete if its closed span equals to H. We say that $\{f_n\}$ is a Bessel sequence if $\sum_{n=1}^{\infty} |\langle f, f_n \rangle|^2 < \infty$ for every element $f \in H$, and that the sequence $\{f_n\}$ is a Riesz-Fischer sequence if the moment problem $\langle f, f_n \rangle = c_n$ ($n = 1, 2, 3, \cdots$) admits at least one solution $f \in H$ whenever $\{c_n\} \in \ell^2$.

A basis $\{f_n\}$ of Hilbert space is called a Riesz basis if it is obtained from an orthonormal basis by means of a bounded linear invertible operator. Two sequences of elements $\{f_n\}$ and $\{g_n\}$ from the Hilbert space H are called quadratically close if $\sum_{n=1}^{\infty} \| f_n - g_n \|^2 < \infty$. A sequence $\{\lambda_n\}$ of real or complex numbers is said to be separated if for some positive number ϵ, $|\lambda_n - \lambda_m| \geqslant \epsilon$ whenever $n \neq m$. A sequence $\{f_n\}$ is called ω-linearly independent if the equality $\sum_{n=1}^{\infty} c_n f_n = 0$ is possible only for $c_n = 0$ $(n \geqslant 1)$.

Next, we need the following lemmas to get our main results.

Lemma 6.1.1 (1) The sequence $\{f_n\}$ is a Bessel sequence with bound M if and only if the inequality

$$\| \sum_n c_n f_n \|^2 \leqslant M \sum_n |c_n|^2 \qquad (6.1.2)$$

holds for every finite system $\{c_n\}$ of complex numbers.

(2) The sequence $\{f_n\}$ is a Riesz-Fischer sequence with bound m if and only if the inequality

$$m \sum_n |c_n|^2 \leqslant \| \sum_n c_n f_n \|^2 \qquad (6.1.3)$$

holds for every finite system $\{c_n\}$ of complex numbers.

Lemma 6.1.2 Let two sequences $\{f_n\}$ and $\{g_n\}$ be quadratically close, $\{f_n\}$ be a Riesz basis in H.

(1) If the sequence $\{g_n\}$ is ω-linearly independent, then $\{g_n\}$ is a Riesz basis in H.
(2) If the sequence $\{g_n\}$ is complete in H, then $\{g_n\}$ is ω-linearly independent.

6.1.2 Riesz bases generated by sines and cosines

Using Lemma 6.1.1 and Lemma 6.1.2, we obtain the following lemmas.

Lemma 6.1.3 If $\{\cos(\lambda_n x)\} \cup \{\sin(\tilde{\lambda}_n x)\}$ is a Riesz-Fischer sequence in $L^2[0, \pi]$ with real λ_n and $\tilde{\lambda}_n$, then the sequences $\{\lambda_n\}$ and $\{\tilde{\lambda}_n\}$ are separated respectively.

Proof. Let m be a lower bound of $\{\cos(\lambda_n x)\} \cup \{\sin(\tilde{\lambda}_n x)\}$. With $c_m = 1, c_k = -1$ and $c_n = 0, d_n = 0$, it follows from (6.1.3) that

$$\sqrt{2m} \leqslant \| \cos(\lambda_m x) - \cos(\lambda_k x) \| . \qquad (6.1.4)$$

On the other hand,

$$\| \cos(\lambda_m x) - \cos(\lambda_k x) \|^2 = \int_0^\pi |\cos(\lambda_m x) - \cos(\lambda_k x)|^2 \mathrm{d}x$$

$$\leqslant \int_0^\pi |\lambda_m - \lambda_k|^2 x^2 \mathrm{d}x \qquad (6.1.5)$$

$$= \frac{\pi^3}{3} |\lambda_m - \lambda_k|^2.$$

Thus $\{\lambda_n\}$ is separated by definition.

Similarly, setting $d_m = 1$, $d_k = -1$ and $c_n = 0$, $d_n = 0$ in (6.1.3), we also have that $\{\widetilde{\lambda}_n\}$ is separated. □

Lemma 6.1.4 Let $\{\lambda_n\} \cup \{\widetilde{\lambda}_n\}$ and $\{\mu_n\} \cup \{\widetilde{\mu}_n\}$, $n \in \mathbb{N}$ be two sequences of nonnegative real numbers such that $\lambda_m \neq \lambda_k$, $\widetilde{\lambda}_m \neq \widetilde{\lambda}_k$, $\mu_m \neq \mu_k$, $\widetilde{\mu}_m \neq \widetilde{\mu}_k$ for all $m \neq k$ and

$$\sum_{n=1}^{\infty}(\lambda_n - \mu_n)^2 + \sum_{n=1}^{\infty}(\widetilde{\lambda}_n - \widetilde{\mu}_n)^2 < \infty. \tag{6.1.6}$$

Then $\{\cos(\lambda_n x)\} \cup \{\sin(\widetilde{\lambda}_n x)\}$ is a Riesz basis in $L^2[0,\pi]$ if and only if $\{\cos(\mu_n x)\} \cup \{\sin(\widetilde{\mu}_n x)\}$ is a Riesz basis in $L^2[0,\pi]$.

Proof. Let $f_n(x) = \cos(\lambda_n x)$, $\widetilde{f}_n(x) = \sin(\widetilde{\lambda}_n x)$ and $g_n(x) = \cos(\mu_n x)$, $\widetilde{g}_n(x) = \sin(\widetilde{\mu}_n x)$. Suppose that $\{f_n\} \cup \{\widetilde{f}_n\}$ is a Riesz basis in $L^2[0,\pi]$. By Lemma 6.1.3, we find that the sequences $\{\lambda_n\}$ and $\{\widetilde{\lambda}_n\}$ are separated respectively. Using (6.1.6), we get that the sequences $\{\mu_n\}$ and $\{\widetilde{\mu}_n\}$ are also separated respectively. Therefore, we can assume

$$\begin{aligned} 0 \leqslant \mu_1 < \mu_2 < \mu_3 < \cdots, \\ 0 \leqslant \widetilde{\mu}_1 < \widetilde{\mu}_2 < \widetilde{\mu}_3 < \cdots \end{aligned} \tag{6.1.7}$$

and there is a positive ϵ such that $\mu_n \geqslant n\epsilon$ and $\widetilde{\mu}_n \geqslant n\epsilon$ for all $n \in \mathbb{N}$. Since

$$\begin{aligned} \|f_n - g_n\| \leqslant \pi|\lambda_n - \mu_n|, \\ \|\widetilde{f}_n - \widetilde{g}_n\| \leqslant \pi|\widetilde{\lambda}_n - \widetilde{\mu}_n|, \end{aligned} \tag{6.1.8}$$

we obtain that

$$\begin{aligned} &\sum_{n=1}^{\infty} \|f_n - g_n\|^2 + \sum_{n=1}^{\infty} \|\widetilde{f}_n - \widetilde{g}_n\|^2 \\ &\leqslant \pi^3 \left(\sum_{n=1}^{\infty}|\lambda_n - \mu_n|^2 + \sum_{n=1}^{\infty}|\widetilde{\lambda}_n - \widetilde{\mu}_n|^2\right) < \infty, \end{aligned} \tag{6.1.9}$$

thus two sequences $\{f_n\} \cup \{\widetilde{f}_n\}$ and $\{g_n\} \cup \{\widetilde{g}_n\}$ are quadratically close. In particular, $\{f_n - g_n\} \cup \{\widetilde{f}_n - \widetilde{g}_n\}$ is a Bessel sequence.

We can define a bounded linear operator

$$\mathrm{T}\left(\sum_{n=1}^{\infty} c_n f_n + \sum_{n=1}^{\infty} d_n \widetilde{f}_n\right) = \sum_{n=1}^{\infty} c_n(f_n - g_n) + \sum_{n=1}^{\infty} d_n(\widetilde{f}_n - \widetilde{g}_n) \tag{6.1.10}$$

in $L^2[0, \pi]$, as $\{f_n\} \cup \{\widetilde{f}_n\}$ is a Riesz basis. From (6.1.9), we have that T is a Hilbert-Schmidt operator. Furthermore, by Lemma 6.1.2, it is sufficient to prove that 1 is a regular point of T in order to prove that $\{g_n\} \cup \{\widetilde{g}_n\}$ is a Riesz basis.

Assume that 1 is not a regular point of T. By the compactness of T, $I - T$ is not one to one. i.e., there exists a sequence $\{c_n\} \cup \{d_n\} \in \ell^2$, not identically zero, such that

$$\sum_{n=1}^{\infty} c_n g_n + \sum_{n=1}^{\infty} d_n \widetilde{g}_n = 0. \tag{6.1.11}$$

Let $\lambda \in \mathbb{C}$ such that $\lambda \neq \pm \mu_n, \pm \widetilde{\mu}_n$ for all $n \in \mathbb{N}$. Then, the series

$$g(x) = \sum_{n=1}^{\infty} \frac{c_n}{\mu_n^2 - \lambda^2} g_n(x) + \sum_{n=1}^{\infty} \frac{d_n}{\widetilde{\mu}_n^2 - \lambda^2} \widetilde{g}_n(x) \tag{6.1.12}$$

is convergent uniformly on $[0, \pi]$. Similarly,

$$\begin{aligned} g'(x) &= \sum_{n=1}^{\infty} \frac{c_n}{\mu_n^2 - \lambda^2} g'_n(x) + \sum_{n=1}^{\infty} \frac{d_n}{\widetilde{\mu}_n^2 - \lambda^2} \widetilde{g}'_n(x) \\ &= -\sum_{n=1}^{\infty} \frac{c_n \mu_n}{\mu_n^2 - \lambda^2} \sin(\mu_n x) + \sum_{n=1}^{\infty} \frac{d_n \widetilde{\mu}_n}{\widetilde{\mu}_n^2 - \lambda^2} \cos(\widetilde{\mu}_n x) \end{aligned} \tag{6.1.13}$$

also converges uniformly on $[0, \pi]$. Because of

$$\begin{aligned} g''_n(x) &= -\mu_n^2 g_n(x), \\ \widetilde{g}''_n(x) &= -\widetilde{\mu}_n^2 \widetilde{g}_n(x), \end{aligned} \tag{6.1.14}$$

we can deduce that

$$\begin{aligned} &\sum_{n=1}^{m} \frac{c_n}{\mu_n^2 - \lambda^2} g''_n(x) + \sum_{n=1}^{m} \frac{d_n}{\widetilde{\mu}_n^2 - \lambda^2} \widetilde{g}''_n(x) \\ &= -\sum_{n=1}^{m} \frac{c_n \mu_n^2}{\mu_n^2 - \lambda^2} g_n(x) - \sum_{n=1}^{m} \frac{c_n \widetilde{\mu}_n^2}{\widetilde{\mu}_n^2 - \lambda^2} \widetilde{g}_n(x) \\ &= -\left(\sum_{n=1}^{m} c_n g_n(x) + \sum_{n=1}^{m} d_n \widetilde{g}_n(x)\right) - \\ &\quad \lambda^2 \left(\sum_{n=1}^{m} \frac{c_n}{\mu_n^2 - \lambda^2} g_n(x) + \sum_{n=1}^{m} \frac{d_n}{\widetilde{\mu}_n^2 - \lambda^2} \widetilde{g}_n(x)\right). \end{aligned} \tag{6.1.15}$$

When $m \to \infty$, the sequence on the right-hand side of (6.1.15) converges to $-\lambda^2 g(x)$ in $L^2[0, \pi]$. This shows that $g(x)$ is twice differentiable and $g''(x) = -\lambda^2 g(x)$ for all $x \in [0, \pi]$. Due to

$$g(0) = \sum_{n=1}^{\infty} \frac{c_n}{\mu_n^2 - \lambda^2},$$
$$g'(0) = \sum_{n=1}^{\infty} \frac{d_n \widetilde{\mu}_n}{\widetilde{\mu}_n^2 - \lambda^2}, \qquad (6.1.16)$$

we obtain that
$$g(x) = u(\lambda) \cos(\lambda x) + v(\lambda) \sin(\lambda x), \qquad (6.1.17)$$

where $u(\lambda) = g(0)$, $v(\lambda) = \lambda^{-1} g'(0)$. The functions $u(\lambda)$ and $v(\lambda)$ are meromorphic and not identically zero, respectively. Thus it has at most countably many zeros. If $u(\lambda)v(\lambda) \neq 0$, by (6.1.12) and (6.1.17), we have that $\{\cos(\lambda x)\} \cup \{\sin(\lambda x)\}$ is in the closed linear span of $\{\cos(\mu_n x)\} \cup \{\sin(\widetilde{\mu}_n x)\}$. Owing to $\{\cos(\lambda x)\} \cup \{\sin(\lambda x)\}$ is continuous about (x, λ), we get that $\{\cos(\lambda x)\} \cup \{\sin(\lambda x)\}$ is in the closed linear span of $\{\cos(\mu_n x)\} \cup \{\sin(\widetilde{\mu}_n x)\}$ for all $\lambda \in \mathbb{C}$. It follows that $\{\sin(nx)\}$, $n \in \mathbb{N}$, is in the closed linear span of $\{g_n(x)\} \cup \{\widetilde{g}_n(x)\}$, so $\{g_n(x)\} \cup \{\widetilde{g}_n(x)\}$ is complete in $L^2[0, \pi]$. Hence $R(I - T)$ is dense in $L^2[0, \pi]$. Using the fact that T is compact, we have that $R(I - T) = L^2[0, \pi]$ and $I - T$ is one to one, this contradicts the assumption.

Similarly, assume that $\{g_n(x)\} \cup \{\widetilde{g}_n(x)\}$ is a Riesz basis in $L^2[0, \pi]$, then $\{f_n(x)\} \cup \{\widetilde{f}_n(x)\}$ is also a Riesz basis in $L^2[0, \pi]$. □

Now we shall introduce our main results.

Theorem 6.1.1 (1) The sequence $\{1\} \cup \{\sin(2nx)\} \cup \{\cos(2nx)\}$ is an orthonormal basis and Riesz basis in $L^2[0, \pi]$.

(2) The sequence $\{\sin((2n-1)x)\} \cup \{\cos((2n-1)x)\}$ is an orthonormal basis and Riesz basis in $L^2[0, \pi]$.

Proof. (1) Suppose that $f(x) \in L^2[0, \pi]$ satisfies
$$\int_0^\pi f(x) \cdot 1 \mathrm{d}x = 0,$$
$$\int_0^\pi f(x) \cos(2nx) \mathrm{d}x = 0, \qquad (6.1.18)$$
$$\int_0^\pi f(x) \sin(2nx) \mathrm{d}x = 0.$$

Let $F(x) = \int_0^x f(t) \mathrm{d}t$, integration by parts yields that
$$\int_0^\pi f(x) \cos(2nx) \mathrm{d}x = F(x) \cos(2nx)|_0^\pi + 2n \int_0^\pi F(x) \sin(2nx) \mathrm{d}x = 0. \qquad (6.1.19)$$

Thus
$$\int_0^\pi F(x)\sin(2nx)\,dx = 0. \tag{6.1.20}$$

Setting $t = 2x - \pi$, we obtain

$$\int_0^\pi F(x)\sin(2nx)\,dx = \frac{(-1)^n}{2}\int_{-\pi}^\pi F\left(\frac{t+\pi}{2}\right)\sin(nt)\,dt = 0, \tag{6.1.21}$$

$$\int_0^\pi f(x)\sin(2nx)\,dx = \frac{(-1)^n}{2}\int_{-\pi}^\pi F'\left(\frac{t+\pi}{2}\right)\sin(nt)\,dt = 0. \tag{6.1.22}$$

Combining (6.1.18), (6.1.21) and (6.1.22), we obtain $f(x) \equiv 0$. Therefore, $\{1\} \cup \{\cos(2nx)\} \cup \{\sin(2nx)\}$, $n \in \mathbb{N}$, is complete in $L^2[0,\pi]$. The orthogonality of $\{1\} \cup \{\cos(2nx)\} \cup \{\sin(2nx)\}$, $n \in \mathbb{N}$, will be proved by establishing that $\{\cos(2nx)\}$ and $\{\sin(2mx)\}$ are orthogonal for all $m,n \in \mathbb{N}$, using the fact that $\{1\} \cup \{\cos(2nx)\}$ and $\{1\} \cup \{\sin(2nx)\}$, $n \in \mathbb{N}$, are the orthogonal sequences in $L^2[0,\pi]$, respectively.

It follows from
$$\langle \cos(2nx), \sin(2mx) \rangle = \int_0^\pi \cos(2nx)\sin(2mx)\,dx = 0, \tag{6.1.23}$$

that $\cos(2nx)$ and $\sin(2mx)$ are orthogonal for all $m,n \in \mathbb{N}$. Clearly, it is also a Riesz basis in $L^2[0,\pi]$.

(2) Suppose $f(x) \in L^2[0,\pi]$, such that
$$\begin{aligned}\int_0^\pi f(x)\sin((2n-1)x)\,dx &= 0,\\ \int_0^\pi f(x)\cos((2n-1)x)\,dx &= 0.\end{aligned} \tag{6.1.24}$$

Let $F(x) = \int_0^x f(t)\,dt$. By partial integration,

$$\begin{aligned}&\int_0^\pi f(x)\sin((2n-1)x)\,dx\\ &= -(2n-1)\int_0^\pi F(x)\cos((2n-1)x)\,dx = 0.\end{aligned} \tag{6.1.25}$$

Hence
$$\int_0^\pi F(x)\cos((2n-1)x)\,dx = 0. \tag{6.1.26}$$

Setting $t = 2x - \pi$, we obtain

$$\int_0^\pi F(x)\cos((2n-1)x)\mathrm{d}x$$
$$= \frac{(-1)^n}{2}\int_{-\pi}^\pi F\left(\frac{t+\pi}{2}\right)\sin\left(\left(n-\frac{1}{2}\right)t\right)\mathrm{d}t = 0,$$
(6.1.27)
$$\int_0^\pi f(x)\cos((2n-1)x)\mathrm{d}x$$
$$= \frac{(-1)^n}{2}\int_{-\pi}^\pi F'\left(\frac{t+\pi}{2}\right)\sin\left(\left(n-\frac{1}{2}\right)t\right)\mathrm{d}t = 0.$$

Similarly, using the method in (1), the desired results can be obtained. The proof is completed. □

Theorem 6.1.2 Let $\delta_n \in \ell^2, n \in \mathbb{N}$.

(1) If $\lambda_n = 2n \pm \delta_n$, $\tilde{\lambda}_n = 2n \mp \delta_n$ and $\lambda_m \neq \lambda_k$, $\tilde{\lambda}_m \neq \tilde{\lambda}_k$, where $m \neq k$, then the sequences $\{1\} \cup \{\cos(\lambda_n x)\} \cup \{\sin(\lambda_n x)\}$ and $\{1\} \cup \{\cos(\lambda_n x)\} \cup \{\sin(\tilde{\lambda}_n x)\}$ are the Riesz basis in $L^2[0,\pi]$, respectively.

(2) If $\check{\lambda}_n = (2n-1) \pm \delta_n$, $\hat{\lambda}_n = (2n-1) \mp \delta_n$ and $\check{\lambda}_m \neq \check{\lambda}_k$, $\hat{\lambda}_m \neq \hat{\lambda}_k$, where $m \neq k$, then the sequences $\{\sin(\check{\lambda}_n x)\} \cup \{\cos(\check{\lambda}_n x)\}$ and $\{\sin(\check{\lambda}_n x)\} \cup \{\cos(\hat{\lambda}_n x)\}$ are the Riesz basis in $L^2[0,\pi]$, respectively.

Proof. (1) By the assumptions (1) of Theorem 6.1.2, we have
$$\sum_{n=1}^\infty (\lambda_n - 2n)^2 = \sum_{n=1}^\infty (\delta_n)^2 < \infty,$$
$$\sum_{n=1}^\infty (\tilde{\lambda}_n - 2n)^2 = \sum_{n=1}^\infty (\delta_n)^2 < \infty.$$
(6.1.28)

Therefore,
$$\sum_{n=1}^\infty (\lambda_n - 2n)^2 + \sum_{n=1}^\infty (\tilde{\lambda}_n - 2n)^2 < \infty.$$
(6.1.29)

Hence the result follows from Lemma 6.1.1 and Lemma 6.1.4.

The proof of the second part of this theorem follows in a similar manner. □

6.1.3 Riesz bases associated with the eigenfunctions of S-L problems

We consider the S-L problem

$$-y'' + q(x)y = \lambda^2 y, \quad x \in [0, \pi],$$
$$y(0) = y(\pi) = 0, \tag{6.1.30}$$

where $\lambda \in \mathbb{C}$ and $q(x) \in L^2([0, \pi], \mathbb{R})$.

It is well known that the eigenvalues of the problem (6.1.30) are

$$\lambda_n = n + O\left(\frac{1}{n}\right) \tag{6.1.31}$$

and corresponding normalized eigenfunctions are

$$y_n(x) = \sqrt{\frac{2}{\pi}} \sin(nx) + O\left(\frac{1}{n}\right). \tag{6.1.32}$$

Theorem 6.1.3 Let $u_n(x, q) = g_1(x, \lambda_n)g_2(x, \lambda_n)$, $n \in \mathbb{N}$, where $g_i(x, \lambda_n)$, $i = 1, 2$, are the solutions to (6.1.30) satisfying the initial conditions

$$g_1(0, \lambda, q) = g_2'(0, \lambda, q) = 1;$$
$$g_1'(0, \lambda, q) = g_2(0, \lambda, q) = 0. \tag{6.1.33}$$

Then, for $m, n \in \mathbb{N}$, we have

(1) $\left\langle y_m^2, \dfrac{\mathrm{d}}{\mathrm{d}x} y_n^2 \right\rangle = 0$;

(2) $\left\langle u_m, \dfrac{\mathrm{d}}{\mathrm{d}x} y_n^2 \right\rangle = \dfrac{\pi}{2}\delta_{mn}$;

(3) $\left\langle u_m, \dfrac{\mathrm{d}}{\mathrm{d}x} u_n \right\rangle = 0$.

Proof. (1) Using integration by parts we obtain

$$\left\langle y_m^2, \frac{\mathrm{d}}{\mathrm{d}x} y_n^2 \right\rangle = \frac{1}{2} \int_0^\pi (y_m^2(y_n^2)' - (y_m^2)'y_n^2)\mathrm{d}x$$
$$= \int_0^\pi y_m y_n [y_m, y_n]\mathrm{d}x. \tag{6.1.34}$$

This clearly vanishes for $m = n$. If $m \neq n$, then $\lambda_n \neq \lambda_m$, and we can use

$$[y_m, y_n]' = (\lambda_m - \lambda_n) y_m y_n \tag{6.1.35}$$

to obtain

$$\left\langle y_m^2, \frac{\mathrm{d}}{\mathrm{d}x} y_n^2 \right\rangle = \frac{1}{2(\lambda_m - \lambda_n)} [y_m, y_n]^2 \big|_0^\pi = 0. \tag{6.1.36}$$

(2) Again, integration by parts yields

$$\left\langle u_m, \frac{\mathrm{d}}{\mathrm{d}x} y_n^2 \right\rangle = \frac{1}{2} \int_0^\pi (u_m(y_n^2)' - (u_m)' y_n^2) \mathrm{d}x$$

$$= \frac{1}{2} \int_0^\pi (2g_1 g_2 y_n y_n' - g_1' g_2 y_n^2 - g_1 g_2' y_n^2) \mathrm{d}x \quad (6.1.37)$$

$$= \frac{1}{2} \int_0^\pi (g_2 y_n [g_1, y_n] + g_1 y_n [g_2, y_n]) \mathrm{d}x,$$

where $g_i = g_i(x, \lambda_m)$ $(i = 1, 2)$. If $m \neq n$, we may use

$$[g_i, y_n]' = (\lambda_m - \lambda_n) g_i y_n, \ i = 1, 2 \quad (6.1.38)$$

to obtain

$$\left\langle u_m, \frac{\mathrm{d}}{\mathrm{d}x} y_n^2 \right\rangle = \frac{1}{2(\lambda_m - \lambda_n)} [g_1, y_n][g_2, y_n]\big|_0^\pi = 0. \quad (6.1.39)$$

If $m = n$, then eigenfunction y_m is a multiple of solution g_2, hence $[g_2, y_m] = 0$, and by the Wronskian identity, we get

$$\left\langle u_m, \frac{\mathrm{d}}{\mathrm{d}x} y_m^2 \right\rangle = \frac{1}{2} \int_0^\pi (u_m(y_m^2)' - u_m' y_m^2) \mathrm{d}x$$

$$= \frac{1}{2} \int_0^\pi g_2 y_m [g_1, y_m] \mathrm{d}x$$

$$= \frac{1}{2} \int_0^\pi y_m^2 [g_1, g_2] \mathrm{d}x \quad (6.1.40)$$

$$= \frac{1}{2} \pi.$$

(3) If $m = n$, then the conclusion holds clearly. If $m \neq n$, using the same procedure in (1) we have

$$\left\langle u_m, \frac{\mathrm{d}}{\mathrm{d}x} u_n \right\rangle = \frac{1}{2} \int_0^\pi (u_m(u_n)' - (u_m)' u_n) \mathrm{d}x$$

$$= \frac{1}{2(\lambda_m - \lambda_n)} [g_1(x, \lambda_m), g_1(x, \lambda_n)][g_2(x, \lambda_m), g_2(x, \lambda_n)]\big|_0^\pi \quad (6.1.41)$$

$$= 0.$$

Theorem 6.1.4 For every $q(x) \in L^2([0, \pi], \mathbb{R})$, the sequence $\{1\} \cup \{1 - \pi y_n^2\} \cup \left\{\frac{\pi}{2n} \frac{\mathrm{d}}{\mathrm{d}x} y_n^2\right\}$, $n \in \mathbb{N}$, is a Riesz basis in $L^2[0, \pi]$.

Proof. It is clear that the element $1 - \pi y_n^2$ is not in the closed linear span of $\{1, 1 - \pi y_m^2\}$, $n \neq m$, as

$$\left\langle 1-\pi y_n^2, \frac{\mathrm{d}}{\mathrm{d}x}u_n\right\rangle = \left\langle 1, \frac{\mathrm{d}}{\mathrm{d}x}u_n\right\rangle - \pi\left\langle y_n^2, \frac{\mathrm{d}}{\mathrm{d}x}u_n\right\rangle = \frac{\pi}{2}, \qquad (6.1.42)$$

but

$$\left\langle 1, \frac{\mathrm{d}}{\mathrm{d}x}u_n\right\rangle = \int_0^\pi \mathrm{d}u_n = g_1 g_2|_0^\pi = 0,$$

$$\left\langle 1-\pi y_m^2, \frac{\mathrm{d}}{\mathrm{d}x}u_n\right\rangle = 0, \ m \neq n, \qquad (6.1.43)$$

by Theorem 6.1.3. Hence $\{1\} \cup \{1 - \pi y_n^2\}$ is ω-linearly independent. Similarly, the sequence $\{\frac{\pi}{2n}\frac{\mathrm{d}}{\mathrm{d}x}y_n^2\}$ is ω-linearly independent. It follows from Theorem 6.1.3 that for all $m, n \in \mathbb{N}$,

$$\left\langle 1, \frac{\mathrm{d}}{\mathrm{d}x}y_n^2\right\rangle = 0, \ \left\langle 1-\pi y_m^2, \frac{\mathrm{d}}{\mathrm{d}x}y_n^2\right\rangle = 0. \qquad (6.1.44)$$

Therefore, the two sequences $\{1\}\cup\{1-\pi y_n^2\}$ and $\{\frac{\pi}{2n}\frac{\mathrm{d}}{\mathrm{d}x}y_n^2\}$ are mutually perpendicular. Hence, the sequence $\{1\} \cup \{1 - \pi y_n^2\} \cup \{\frac{\pi}{2n}\frac{\mathrm{d}}{\mathrm{d}x}y_n^2\}$ is ω-linearly independent.

By the expression of $y_n(x)$, we have

$$1 = 1,$$

$$1 - \pi y_n^2 = \cos(2nx) + O\left(\frac{1}{n}\right), \qquad (6.1.45)$$

$$\frac{\pi}{2n}\frac{\mathrm{d}}{\mathrm{d}x}y_n^2 = \sin(2nx) + O\left(\frac{1}{n}\right).$$

Thus the sequence $\{1\}\cup\{1-\pi y_n^2\}\cup\{\frac{\pi}{2n}\frac{\mathrm{d}}{\mathrm{d}x}y_n^2\}$ is quadratically close to the Riesz basis $\{1\} \cup \{\cos(2nx)\} \cup \{\sin(2nx)\}$. The statement follows directly from Lemma 6.1.2. □

6.2 Eigenvalue problems of internal solitary waves

Nonlinear internal solitary waves have attracted great attention in the past decades, both field experiments and remote-sensing observations show that internal solitary waves frequently occur in sea shelves and marginal seas. The EKdV equation is a classical model describing the internal solitary waves. To be understood easily, the derivation process of the EKdV equation will be reviewed below.

6.2.1 The constant coefficient theoretical model

Assuming that the fluid layer is incompressible and inviscid, the two-dimensional governing equations are[131]

$$u_t + uu_x + wu_z = -p_x, \qquad (6.2.1)$$

$$w_t + uw_x + ww_z = -p_z - \rho g, \qquad (6.2.2)$$

$$u_x + w_z = 0, \qquad (6.2.3)$$

$$\rho_t + u\rho_x + w\rho_z = 0, \qquad (6.2.4)$$

where $u(x, z, t)$ and $w(x, z, t)$ are horizontal and vertical velocities, respectively. g is gravity, p is the pressure, the subscripts x, z and t represent the derivatives with respect to variable x, z and t. Take the vertically upward direction as positive. ρ is the density change over the vertical characteristic length.

The vorticity equation can be obtained by eliminating the pressure term from Equations (6.2.1) and (6.2.2)

$$(u_t + uu_x + wu_z = -p_x)_z - (w_t + uw_x + ww_z)_x - g\rho_x = 0. \qquad (6.2.5)$$

Assuming that

$$\rho = \bar{\rho}(z) + \rho'(x, z, t), u(z) = \bar{u}(z) + u'(x, z, t),$$

where $\bar{\rho}(z)$ is the background density, $\rho'(x, z, t)$ is the disturbance density, $\bar{u}(z)$ is the mean flow, $u'(x, z, t)$ is the fluctuation velocity. Combining Equations (6.2.3), (6.2.4) and (6.2.5), the dimensionless flow function governing equations are obtained as follows[132]

$$\mu\frac{\partial}{\partial t}\Phi_{xx} + \frac{\partial}{\partial t}\Phi_{zz} - b_x = \varepsilon\mu J(\Phi, \Phi_{xx}) + \varepsilon J(\Phi, \Phi_{xx}) + \Phi_x\bar{u}_{zz} - \bar{u}(\mu\Phi_{xx} + \Phi zz)_x, \qquad (6.2.6)$$

$$b_t + N^2(z)\Phi_x = \varepsilon J(\Phi, b) - \bar{u}b_x, \qquad (6.2.7)$$

where

$$(u, w) = (\Phi_z, -\Phi_x), b = g\rho', N^2 = -g\frac{\mathrm{d}(\bar{\rho})}{\mathrm{d}z}, J(A, B) = \eta_x B_z - \eta_z B_x,$$

$N(z)$ is the Buoyancy frequency of the background field, ε and μ are dimensionless parameters. The boundary conditions of governing Equations (6.2.6) and (6.2.7) are as follows

$$\Phi(0) = \Phi(1) = 0, b(0) = b(1) = 0.$$

By asymptotic expansion of [133]

$$\Phi \sim \eta(x,t)W(z) + \varepsilon\eta^2(x,t)W^{1,0}(z) + \mu\eta_{xx}(x,t)W^{0,1}(z) +$$
$$\varepsilon^2\eta^3(x,t)W^{2,0}(z) + O(\varepsilon\mu, \mu^2),$$
$$b \sim \eta(x,t)D(z) + \varepsilon\eta^2(x,t)D^{1,0}(z) + \mu\eta_{xx}(x,t)D^{0,1}(z) +$$
$$\varepsilon^2\eta^3(x,t)D^{2,0}(z) + O(\varepsilon\mu, \mu^2),$$

the linear approximation

$$W_{zz} + \left[\frac{N^2}{(c-\bar{u})^2} + \frac{\bar{u}_{zz}}{c-\bar{u}}\right]W = 0, \tag{6.2.8}$$

$$\eta_t = -c\eta_x, \tag{6.2.9}$$

$$D = \frac{N^2}{c-\bar{u}}W \tag{6.2.10}$$

is obtained. Equation (6.2.8) and boundary conditions

$$W(0) = W(1) = 0$$

form a Sturm-Liouville problem, where W represents the amplitude of the vertical movement of the internal wave water particle.

If the controlling equations (6.2.6) and (6.2.7) are accurate to $o(\varepsilon^2)$, get the EKdV equation

$$\eta_t + c\eta_x + \alpha\eta\eta_x + \alpha_1\eta^2\eta_x + \beta\eta_{xxx} = 0,$$

where, c is the longwave phase speed, α is the quadratic nonlinearity coefficient, and α_1 is the cubic nonlinearity coefficient, β is the dispersion coefficient.

6.2.2 The variable coefficient theoretical model

In the actual ocean, along the direction of internal solitary wave propagation, the topography, stratification and background velocity are constantly changing. Therefore, the model coefficients are constantly changing during the propagation process. The variable coefficient EKdV equation can be expressed as[134]

$$\frac{\partial\eta}{\partial t} + (c + \alpha\eta + \alpha_1\eta^2)\frac{\partial\eta}{\partial x} + \beta\frac{\partial^3\eta}{\partial x^3} + \frac{c}{2Q}\frac{dQ}{dx}\eta = 0, \tag{6.2.11}$$

c in (6.2.11) is determined by an eigenvalue problem

$$\frac{d}{dz}\left[(c-\bar{u})^2\frac{dW}{dz}\right] + N^2 W = 0,$$

$$W(0) = W(-H) = 0.$$

The nonlinear coefficient and the dispersion coefficient can be calculated by the following expressions

$$\alpha = \frac{3}{2} \frac{\int_{-H}^{0} (c-\bar{u})^2 \left(\dfrac{\mathrm{d}W}{\mathrm{d}z}\right)^3 \mathrm{d}z}{\int_{-H}^{0} (c-\bar{u}) \left(\dfrac{\mathrm{d}W}{\mathrm{d}z}\right)^2 \mathrm{d}z}, \tag{6.2.12}$$

$$\alpha_1 = \frac{\int_{-H}^{0} \left\{\left[\left(\dfrac{\mathrm{d}W}{\mathrm{d}z}\right)^2 - \alpha^2 \left(\dfrac{\mathrm{d}W}{\mathrm{d}z}\right)^2 + \Pi_1\right] \Pi_2 \right\} \mathrm{d}z}{\int_{-H}^{0} (c-\bar{u}) \left(\dfrac{\mathrm{d}W}{\mathrm{d}z}\right)^2 \mathrm{d}z}, \tag{6.2.13}$$

$$\beta = \frac{1}{2} \frac{\int_{-H}^{0} (c-\bar{u})^2 W^2 \mathrm{d}z}{\int_{-H}^{0} (c-\bar{u}) \left(\dfrac{\mathrm{d}W}{\mathrm{d}z}\right)^2 \mathrm{d}z}, \tag{6.2.14}$$

$$Q = \frac{c^2 \int_{-H}^{0} (c-\bar{u}) \left(\dfrac{\mathrm{d}W}{\mathrm{d}z}\right)^2 \mathrm{d}z}{c_0^2 \int_{-H}^{0} (c_0-\bar{u}_0) \left(\dfrac{\mathrm{d}W_0}{\mathrm{d}z}\right)^2 \mathrm{d}z}, \tag{6.2.15}$$

where

$$\Pi_1 = \alpha(c-\bar{u})\left[5\left(\frac{\mathrm{d}W}{\mathrm{d}z}\right)^2 - 4\frac{\mathrm{d}T}{\mathrm{d}z}\right]\frac{\mathrm{d}W}{\mathrm{d}z},$$

$$\Pi_2 = 3(c-\bar{u})^2 \left[3\frac{\mathrm{d}T}{\mathrm{d}z} - 2\left(\frac{\mathrm{d}W}{\mathrm{d}z}\right)^2\right].$$

T is determined by an inhomogeneous eigenvalue problem

$$\frac{\mathrm{d}}{\mathrm{d}z}\left[(c-\bar{u})^2 \frac{\mathrm{d}T}{\mathrm{d}z}\right] + N^2 T$$

$$= -\alpha \frac{\mathrm{d}}{\mathrm{d}z}\left[(c-\bar{u})\frac{\mathrm{d}W}{\mathrm{d}z}\right] + \frac{3}{2}\frac{\mathrm{d}}{\mathrm{d}z}\left[(c-\bar{u})^2 \left(\frac{\mathrm{d}W}{\mathrm{d}z}\right)^2\right],$$

$$T(0) = T(-H) = 0.$$

The subscript "0" in (6.2.15) indicates the value at any fixed point x. The isopycnal vertical displacement can be expressed as

$$\xi(x, z, t) = \eta(x, t)W(z) + \eta^2(x, t)T(z),$$

where $\eta(x, t)$ is the vertical displacement at depth $z_{\max}$, $z_{\max}$ is the depth at the maximum value of the normalized linear modal function. The stream function is as follows

$$\Phi(x, z, t) = c\eta(x, t)W(z) + c\eta^2(x, t)T(z).$$

6.2.3 Eigenvalue problems

The equations of free internal wave without considering the influence of background flow field are as follows[135]

$$\begin{cases} \dfrac{d^2W}{dz^2} + (N^2/g)\dfrac{dW}{dz} + k(k+\beta/\omega)(N^2-\omega^2)/(\omega^2-f^2)W = 0, \\ W(-H) = 0, \\ \dfrac{dW(0)}{dz} - gk^2/(\omega^2-f^2)W(0) = 0, \end{cases} \quad (6.2.16)$$

where ω represents the frequency; k is the wave number in the corresponding horizontal direction; f is the Coriolis parameter, $\beta = df/dy$; H is the ocean depth, which is taken as a constant.

Taking $\beta = 0$, then (6.2.16) is simplified as

$$\begin{cases} \dfrac{d^2W}{dz^2} + (N^2/g)\dfrac{dW}{dz} + k^2(N^2-\omega^2)/(\omega^2-f^2)W = 0, \\ W(-H) = 0, \\ \dfrac{dW(0)}{dz} - gk^2/(\omega^2-f^2)W(0) = 0. \end{cases} \quad (6.2.17)$$

Further, if there is no fluctuation in the sea surface, then (6.2.17) is simplified as

$$\begin{cases} \dfrac{d^2W}{dz^2} + (N^2/g)\dfrac{dW}{dz} + k^2(N^2-\omega^2)/(\omega^2-f^2)W = 0, \\ W(-H) = 0, \ W(0) = 0, \end{cases} \quad (6.2.18)$$

By introducing Boussinesq approximation, the corresponding internal solitary wave equation can be further simplified as follows

$$\begin{cases} \dfrac{d^2W}{dz^2} + k^2(N^2-\omega^2)/(\omega^2-f^2)W = 0, \\ W(-H) = 0, \ W(0) = 0. \end{cases} \quad (6.2.19)$$

This is the Fjelsted internal wave equation in ocean dynamics. Furthermore, under the condition of large-scale waves, if the geostrophic effect is ignored, the corresponding internal wave equation can be simplified as follows:

$$\begin{cases} \dfrac{d^2 W}{dz^2} + N^2/c^2 W = 0, \\ W(-H) = 0, \ W(0) = 0. \end{cases} \quad (6.2.20)$$

6.2.4 Applications of models

Assume that the bathymetry used for the calculation is[7]

$$H = \begin{cases} 560, \ x \leqslant 2500, \\ 560 + 105\left(\cos\pi\dfrac{x-2500}{75000} - 1\right), \ x \leqslant 40000, \\ 455.63 - 23.63\dfrac{x-40000}{14000}, \ x \leqslant 52500, \\ 404.53, \ x \geqslant 52500, \end{cases} \quad (6.2.21)$$

as shown in the Figure 6.2.3(f). The density ρ and the Buoyancy frequency N follows (shown in Figure 6.2.1)

$$N^2(z) = -\dfrac{g}{\rho_0}\dfrac{d\rho}{dz}, \quad (6.2.22)$$

where ρ_0 is the potential density.

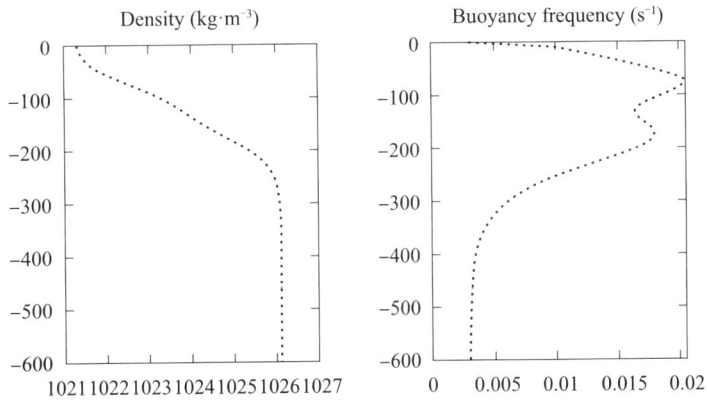

Figure 6.2.1 The Density and Buoyancy frequency

The model functions can be obtained by using (6.2.20). Taking stations 37750 m and 42750 m as examples, the variations of the model functions with the bathymetry are shown in Figure 6.2.2.

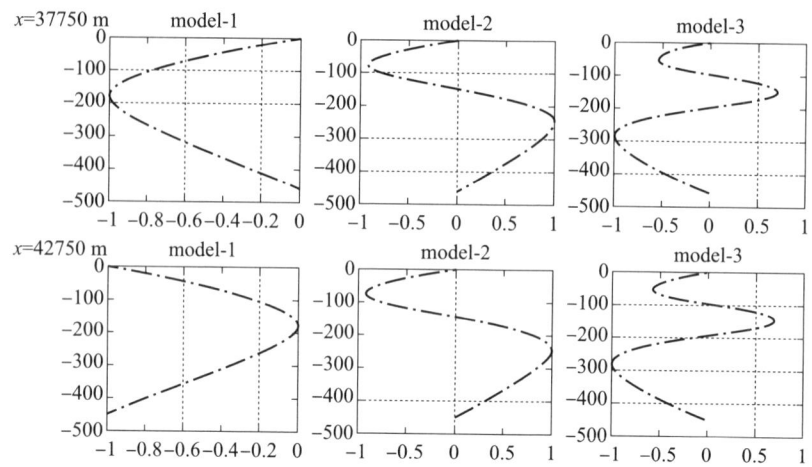

Figure 6.2.2 The variations of the model functions with the depth of water

The coefficients of EKdV modal (6.2.11) can be obtained by using (6.2.12) and (6.2.13). The variations of the model functions with the bathymetry are shown in Figure 6.2.3.

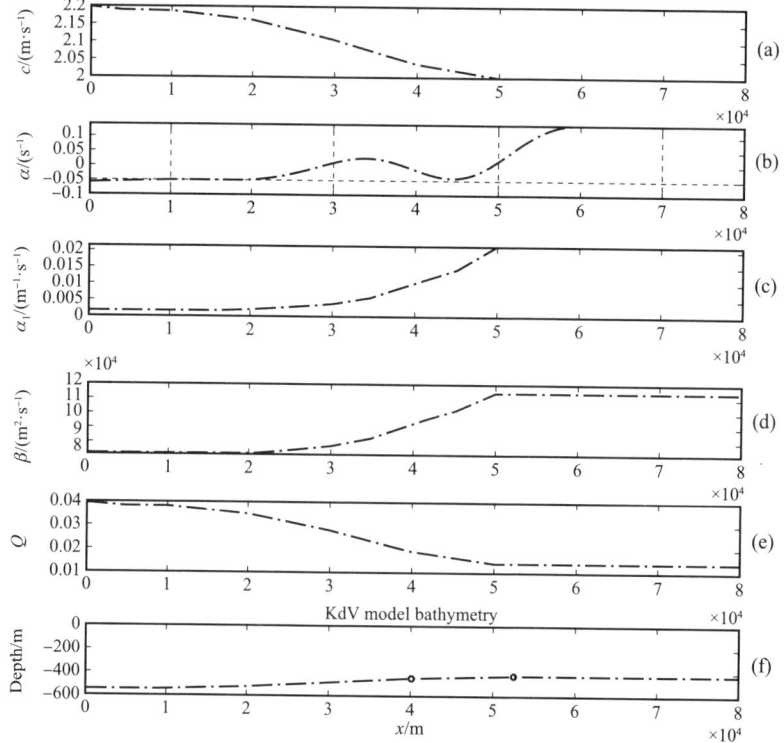

Figure 6.2.3 The variations of coefficients with the bathymetry

With the change of the bathymetry and other conditions, the sign of the corresponding nonlinear coefficient α will change from a positive sign to a negative sign, or from a negative sign to a positive sign, resulting in the change of the polarity of the internal solitary wave, that is, from a concave type to a convex type, or from a convex type to a concave type.

Appendix A Fundamentals Sturm-Liouville problems

This appendix originated from the related sections in [3].

A.1 Classes of Sturm-Liouville problems

A Sturm-Liouville equation(SLE) is a differential equation of the form

$$-(py')' + qy = \lambda w y \tag{A.1.1}$$

on $J = -\infty \leqslant a < b \leqslant +\infty$, where p is called the leading coefficient, q is called the potential, and w is called the density or the weight function, satisfying

$$1/p, q, w \in L_{\text{loc}}(J, \mathbb{C}), \tag{A.1.2}$$

w is not almost everywhere (a.e.) 0 on J, and $\lambda \in \mathbb{C}$ is the so-called spectral parameter. Note that (A.1.2) implies that $p \neq 0$ a.e. on J. By a solution to (A.1.1) we mean a function y on J such that y and py' are absolutely continuous in all compact subintervals of J and satisfy (A.1.1) a.e.. For any function y, the function py' is called the quasi-derivative of y relative to (A.1.1).

Example A.1.1 The SLE

$$-y'' = \lambda y \tag{A.1.3}$$

on J is usually called the Fourier equation on J. If $\lambda \neq 0$, then any solution to (A.1.3) can be written as

$$y(t) = c_1 \cos(\sqrt{\lambda} t) + c_2 \sin(\sqrt{\lambda} t) \tag{A.1.4}$$

when $\lambda = 0$, the solutions to (A.1.3) have the form

$$y(t) = c_1 + c_2 t, \tag{A.1.5}$$

here c_1 and c_2 are complex constants.

Example A.1.2 Let $t_* \in J$. The SLE

$$-y''(t) = \lambda \, \text{sgn}\,(t - t_*)\, y(t) \tag{A.1.6}$$

on J will be called the modified Fourier equation on J with jump at t_*. For any λ, (A.1.6) has a solution given by

$$y(t) = \begin{cases} \cosh(\sqrt{\lambda}t), & \text{if } t \in (a, t_*] \\ \cosh(\sqrt{\lambda}t_*)\cos((\sqrt{\lambda}t(t-t_*))+ \\ \sinh(\sqrt{\lambda}t_*)\sin((\sqrt{\lambda}(t-t_*)), & \text{if } t \in (t_*, b). \end{cases} \quad (A.1.7)$$

Example A.1.3 Let

$$p(t) = w(t) = \begin{cases} 1, & \text{if } t \in (0,1) \\ 2, & \text{if } t \in (1,2) \end{cases} \quad (A.1.8)$$

Then, the SLE

$$-(py')' = \lambda wy \quad (A.1.9)$$

in $(0,2)$ is closely related to the Fourier equation in $(0,2)$. However, the solution $\cos(\sqrt{\lambda}t)$ of the Fourier equation is not a solution to (A.1.9) in general:

$$p(t)(\cos(\sqrt{\lambda}t))' = \begin{cases} -\sqrt{\lambda}\sin(\sqrt{\lambda}t), & \text{if } t \in (0,1], \\ -2\sqrt{\lambda}\sin(\sqrt{\lambda}t), & \text{if } t \in (1,2) \end{cases} \quad (A.1.10)$$

is not continuous at $t = 1$ in general. A related solution to (A.1.9) is

$$y(t) = \begin{cases} \cos(\sqrt{\lambda}t), & \text{if } t \in (0,1], \\ \cos\sqrt{\lambda}\cos\left(\sqrt{\lambda}(t-1) - \dfrac{1}{2}\sin\sqrt{\lambda}\sin\sqrt{\lambda}(t-1)\right), & \text{if } t \in (1,2). \end{cases}$$
$$(A.1.11)$$

Let $r = 1/p$ and $\boldsymbol{Y} = (y\ py')^T$, then the SLE (A.1.1) is equivalent to the first-order linear differential equation

$$\boldsymbol{Y}' = \begin{pmatrix} 0 & r \\ q - \lambda w & 0 \end{pmatrix} \boldsymbol{Y} \quad (A.1.12)$$

on J, which will be called the matrix form of (A.1.1). Note that if

$$r, q, w \in L_{\text{loc}}(J, \mathbb{C}), \quad (A.1.13)$$

then (A.1.12) can be written as an SLE if and only if

$$r \neq 0 \text{ a.e. on } J. \quad (A.1.14)$$

Theorem A.1.1 If (A.1.2) holds, then for any $t_0 \in J$ and any $c_0, c_1 \in \mathbb{C}$, the initial value problem

$$-(py')' + qy = \lambda wy, \quad y(t_0) = c_0, \quad (py')(t_0) = c_1 \tag{A.1.15}$$

has a unique solution y in the whole interval J.

Definition A.1.1 The left endpoint a (finite or infinite) of the SLE (A.1.1) is said to be regular if, in addition to (A.1.2),

$$1/p, q, w \in L((a, b_*), \mathbb{C}) \tag{A.1.16}$$

for some (and hence all) $b_* \in J$; otherwise, a is said to be singular. Similar definitions are made for the right endpoint b. The SLE (A.1.1) is said to be regular if both of its endpoints are regular, i.e., if actually

$$1/p, q, w \in L(J, \mathbb{C}); \tag{A.1.17}$$

otherwise, i.e., if only (A.1.2) holds, but not (A.1.17), the SLE (A.1.1) is said to be singular.

For example, the Fourier equation (or the modified Fourier equation) on J is regular if and only if J is finite.

Theorem A.1.2 Assume that (A.1.2) holds.

(1) If (A.1.1) is regular at its left endpoint a, then: first, any solution y to (A.1.1) and its quasi-derivative py' can be extended to a by continuity to become absolutely continuous on the interval $[a, b)$, which is the definition of $y(a)$ and $(py')(a)$ that we will take from now on; second, for any $t_0 \in [a, b)$ and any $c_0, c_1 \in \mathbb{C}$, there is a unique solution y to (A.1.1) such that $y(t_0) = c_0$ and $(py')(t_0) = c_1$.

(2) We have similar results for the right endpoint b of (A.1.1) and a similar definition of $y(b)$ and $(py')(b)$.

In particular, if (A.1.1) is regular, then: first, any solution to (A.1.1) can be extended to both a and b by continuity to become absolutely continuous in the interval $[a, b]$; second, for any $t_0 \in [a, b]$ and any $c_0, c_1 \in \mathbb{C}$, there is a unique solution y to (A.1.1) such that $y(t_0) = c_0$ and $(py')(t_0) = c_1$.

Remark A.1.1 If (A.1.1) is regular and y is a solution to (A.1.1), then y is bounded on $[a, b]$ and hence $y\bar{y}w \in L(J, \mathbb{C})$

From now on, we will denote by $y^{[1]}$ the quasi-derivative py' of y relative to (A.1.1). A regular boundary condition is a relation specified by an algebraic equation of the form

$$(\boldsymbol{A} \mid \boldsymbol{B}) \begin{pmatrix} y(a) \\ y^{[1]}(a) \\ y(b) \\ y^{[1]}(b) \end{pmatrix} = 0, \quad \text{i.e.} \quad \boldsymbol{A}Y(a) + \boldsymbol{B}Y(b) = 0, \tag{A.1.18}$$

where
$$A, B \in M_{2,2}^{\mathbb{C}} \quad \text{such that} \quad (A \mid B) \in M_{2,4}^{\mathbb{C}*}. \tag{A.1.19}$$

Note that equivalent algebraic equations of this form define the same boundary condition. Denote by $[A \mid B]$ the equivalence class containing $(A \mid B)$ of this equivalence relation on $M_{2,4}^{\mathbb{C}*}$.

A regular Sturm-Liouville problem (SLP) consists of a regular SLE and a regular boundary condition. In the rest of this book, we will always assume that (A.1.17) holds, i.e., we will only consider regular SLE and regular SLP and hence, omit the word "regular" from all statements.

Each value of λ for which the SLE (A.1.1) has a non-trivial solution satisfying the boundary condition (A.1.18) is called an eigenvalue of the SLP consisting of (A.1.1) and (A.1.18) and such a solution is called an eigenfunction for this eigenvalue. The vector space spanned by the eigenfunctions for an eigenvalue is the eigenspace for the eigenvalue, while the dimension of the eigenspace is called the geometric multiplicity of the eigenvalue. By Corollary C.2.1, the geometric multiplicity of any eigenvalue is either 1 or 2.

Example A.1.4 Let J be finite. The SLP consisting of the Fourier equation (A.1.3) on J and the so-called Dirichlet boundary condition
$$\begin{pmatrix} 1 & 0 & 0 & 0 \\ 0 & 0 & -1 & 0 \end{pmatrix} \begin{pmatrix} Y(a) \\ Y(b) \end{pmatrix} = \mathbf{0}, \text{ i.e., } y(a) = 0 = y(b) \tag{A.1.20}$$

has eigenvalues
$$\lambda_n = \frac{n^2 \pi^2}{(b-a)^2}, \quad n \in \mathbb{N}. \tag{A.1.21}$$

For each $n \in \mathbb{N}$, λ_n has geometric multiplicity 1 and an eigenfunction for λ_n is $\sin(\sqrt{\lambda_n}(t-a))$. The SLP consisting of the Fourier equation on J and the so-called periodic boundary condition
$$\begin{pmatrix} 1 & 0 & -1 & 0 \\ 0 & 1 & 0 & -1 \end{pmatrix} \begin{pmatrix} Y(a) \\ Y(b) \end{pmatrix} = \mathbf{0}, \quad \text{i.e.} \quad Y(a) = Y(b) \tag{A.1.22}$$

has eigenvalues
$$\lambda_n = \frac{4(n-1)^2 \pi^2}{(b-a)^2}, \quad n \in \mathbb{N}. \tag{A.1.23}$$

In this case, $\lambda_1 = 0$ has geometric multiplicity 1 and an eigenfunction for λ_1 is the constant function 1, while for each $n \in \mathbb{N}$ satisfying $n \geqslant 2$, λ_n has geometric multiplicity 2 and the corresponding non-trivial solutions, i.e., $c_1 \cos\left(\sqrt{\lambda_n}(t-a)\right) + c_2 \sin\left(\sqrt{\lambda_n}(t-a)\right)$, where c_1 and c_2 are constants and not both 0, of the Fourier equation are the eigenfunctions for λ_n.

The easiest classification of boundary conditions is the one determined by the reality of a coefficient matrix: a boundary condition is real if it can be defined by an algebraic equation with a real coefficient matrix; otherwise, it is non-real.

The second classification is according to the form of a coefficient matrix. Boundary conditions that can be written into the form

$$\begin{pmatrix} a_{11} & a_{12} & 0 & 0 \\ 0 & 0 & b_{21} & b_{22} \end{pmatrix} \begin{pmatrix} \mathbf{Y}(a) \\ \mathbf{Y}(b) \end{pmatrix} = \mathbf{0} \tag{A.1.24}$$

are called separated ones. Among the separated boundary conditions are the Dirichlet boundary condition (A.1.20), the Dirichlet-Neumann boundary condition

$$\begin{pmatrix} 1 & 0 & 0 & 0 \\ 0 & 0 & 0 & -1 \end{pmatrix} \begin{pmatrix} \mathbf{Y}(a) \\ \mathbf{Y}(b) \end{pmatrix} = \mathbf{0}, \quad \text{i.e., } y(a) = 0 = y^{[1]}(b), \tag{A.1.25}$$

the Neumann-Dirichlet boundary condition

$$\begin{pmatrix} 0 & 1 & 0 & 0 \\ 0 & 0 & -1 & 0 \end{pmatrix} \begin{pmatrix} \mathbf{Y}(a) \\ \mathbf{Y}(b) \end{pmatrix} = \mathbf{0}, \text{ i.e., } y^{[1]}(a) = 0 = y(b), \tag{A.1.26}$$

the Neumann boundary condition

$$\begin{pmatrix} 0 & 1 & 0 & 0 \\ 0 & 0 & 0 & -1 \end{pmatrix} \begin{pmatrix} \mathbf{Y}(a) \\ \mathbf{Y}(b) \end{pmatrix} = \mathbf{0}, \text{ i.e., } y^{[1]}(a) = 0 = y^{[1]}(b). \tag{A.1.27}$$

The boundary condition (A.1.18) is separated if and only if $\det \mathbf{A} = \det \mathbf{B} = 0$. Any eigenvalue for a separated boundary condition has geometric multiplicity 1.

A boundary condition that is not separated and not one of the degenerated boundary conditions (actually the trivial initial conditions)

$$\begin{pmatrix} 1 & 0 & 0 & 0 \\ 0 & 1 & 0 & 0 \end{pmatrix} \begin{pmatrix} \mathbf{Y}(a) \\ \mathbf{Y}(b) \end{pmatrix} = \mathbf{0}, \text{ i.e., } y(a) = 0 = y^{[1]}(a) \tag{A.1.28}$$

and

$$\begin{pmatrix} 0 & 0 & -1 & 0 \\ 0 & 0 & 0 & -1 \end{pmatrix} \begin{pmatrix} \mathbf{Y}(a) \\ \mathbf{Y}(b) \end{pmatrix} = \mathbf{0}, \text{ i.e., } y(b) = 0 = y^{[1]}(b) \tag{A.1.29}$$

is called a coupled one. Note that there is no eigenvalue for each of the degenerated boundary conditions. The periodic boundary condition (A.1.22) and the semi-periodic boundary condition

$$\begin{pmatrix} 1 & 0 & 1 & 0 \\ 0 & 1 & 0 & 1 \end{pmatrix} \begin{pmatrix} Y(a) \\ Y(b) \end{pmatrix} = 0, \text{ i.e., } Y(a) = -Y(b) \tag{A.1.30}$$

are two important coupled boundary conditions.

The last classification of boundary conditions that we discuss here is closely related to the reality of eigenvalues. The boundary condition (A.1.18) is said to be self-adjoint if

$$A \begin{pmatrix} 0 & 1 \\ -1 & 0 \end{pmatrix} A^* = B \begin{pmatrix} 0 & 1 \\ -1 & 0 \end{pmatrix} B^* \tag{A.1.31}$$

where $A^* = \bar{A}^T$ is the complex conjugate transpose of A; otherwise, the boundary condition is nonself-adjoint. Note that self-adjointness is well-defined: the matrix $(A \mid B)$ in (A.1.18) satisfies (A.1.31) if and only if the coefficient matrix of any equivalent algebraic equation does. The Dirichlet boundary condition, the Dirichlet-Neumann boundary condition, the Neumann-Dirichlet boundary condition, the Neumann boundary condition, the periodic boundary condition and the semi-periodic boundary condition are all self-adjoint.

Lemma A.1.1 (1) If the boundary condition (A.1.18) is self-adjoint, then

$$|\det A| = |\det B| \tag{A.1.32}$$

and hence either (A.1.18) is separated or both A and B are invertible.

(2) A separated boundary condition is self-adjoint if and only if it is real.

(3) A real coupled boundary condition $[A \mid B]$ is self-adjoint if and only if

$$[A \mid B] = [K \mid -I]$$

for some $K \in \mathrm{SL}(2, \mathbb{R})$, where $\mathrm{SL}(2, \mathbb{R})$ is special linear group, the multiplicative group of all 2×2 real matrices with determinant 1.

(4) A coupled boundary condition $[A \mid B]$ is self-adjoint if and only if

$$[A \mid B] = [e^{i\theta} K \mid -I]$$

for some $\theta \subset [0, \pi)$ and $K \subset \mathrm{SL}(2, \mathbb{R})$.

Theorem A.1.3 If p and q are real-valued on J and

$$w > 0 \text{ a.e. on } J, \tag{A.1.33}$$

then the eigenvalues of the Sturm-Liouville problem consisting of (A.1.1) and a self-adjoint boundary condition are all real.

Definition A.1.2 The SLP consisting of (A.1.1) and (A.1.18) is said to be self-adjoint if $p(t)$ and $q(t)$ are real-valued on J, $w > 0$ a.e. on J and (A.1.18) is self-adjoint; otherwise, it is non-self-adjoint.

A.2 Characteristic function

In this section, we introduce the characteristic function, which determines the eigenvalues of SLPs, and prove that it is entire in the spectral parameter.

For each $\lambda \in \mathbb{C}$, let $\varphi_{11}(t, \lambda)$ and $\varphi_{12}(t, \lambda)$ be the solutions to (A.1.1) determined by the initial conditions

$$\varphi_{11}(a, \lambda) = 1. \quad \varphi_{11}^{[1]}(a, \lambda) = 0, \quad \varphi_{12}(a, \lambda) = 0, \quad \varphi_{12}^{[1]}(a, \lambda) = 1. \tag{A.2.1}$$

We will denote $\varphi_{11}^{[1]}$ and $\varphi_{12}^{[1]}$ by φ_{21} and φ_{22}, respectively. Set

$$\boldsymbol{\Phi}(t, \lambda) = \begin{pmatrix} \varphi_{11}(t, \lambda) & \varphi_{12}(t, \lambda) \\ \varphi_{21}(t, \lambda) & \varphi_{22}(t, \lambda) \end{pmatrix}, \quad t \in [a, b], \quad \lambda \in \mathbb{C}. \tag{A.2.2}$$

Then, $\boldsymbol{\Phi}(t, \lambda)$ is the solution of the matrix form (A.1.12) to (A.1.1) satisfying $\boldsymbol{\Phi}(a, \lambda) = \boldsymbol{I}$, where $\boldsymbol{I}$ is the identity matrix. Thus, by Remark of corollary C.2.1, any solution to (A.1.1) is a linear combination of $\varphi_{11}(t, \lambda)$ and $\varphi_{12}(t, \lambda)$. Moreover,

$$\boldsymbol{\Phi}(t, \lambda) \in \text{SL}(2, \mathbb{C}), \quad t \in [a, b], \quad \lambda \in \mathbb{C} \tag{A.2.3}$$

and, if actually p, q and w are real-valued,

$$\boldsymbol{\Phi}(t, \lambda) \in \text{SL}(2, \mathbb{R}), \quad t \in [a, b], \quad \lambda \in \mathbb{R}. \tag{A.2.4}$$

We will call $\boldsymbol{\Phi}$ the fundamental matrix of (A.1.1). For example, when J is finite, the fundamental matrix of the Fourier equation on J is

$$\boldsymbol{\Phi}(t, \lambda) = \begin{pmatrix} \cos f(t, \lambda) & \frac{1}{\sqrt{\lambda}} \sin f(t, \lambda) \\ -\sqrt{\lambda} \sin f(t, \lambda) & \cos f(t, \lambda) \end{pmatrix}, \quad t \in [a, b], \quad \lambda \in \mathbb{C}, \tag{A.2.5}$$

where $f(t, \lambda) = (t-a)\sqrt{\lambda}$. Note that $\frac{1}{\sqrt{\lambda}} \sin((t-a)\sqrt{\lambda})$ always has a natural extension to $\lambda = 0$ using its limit $t - a$ there. The following result says that $\boldsymbol{\Phi}(b, \lambda)$ characterizes the eigenvalues of the SLPs.

Theorem A.2.1 A number $\lambda \in \mathbb{C}$ is an eigenvalue of the S-L problem consisting of (A.1.1) and (A.1.18) if and only if

$$\Delta(\lambda) =: \det(\boldsymbol{A} + \boldsymbol{B}\boldsymbol{\Phi}(b, \lambda)) = 0 \tag{A.2.6}$$

Remark A.2.1 The above proof also shows the following: a number $\lambda_* \in \mathbb{C}$ is an eigenvalue for (A.1.18) of geometric multiplicity 1 if and only if $\boldsymbol{A} + \boldsymbol{B}\boldsymbol{\Phi}(b, \lambda_*)$ has rank 1; in this case, if $(d_1\ d_2)$ is a non-zero row of $\boldsymbol{A} + \boldsymbol{B}\boldsymbol{\Phi}(b, \lambda_*)$, then $d_2\varphi_{11}(t, \lambda_*) - d_1\varphi_{12}(t, \lambda_*)$ is an eigenfunction for λ_*. (When λ_* is an eigenvalue of geometric multiplicity 2, any non-trivial linear combination $c_1\varphi_{11}(t, \lambda_*) + c_2\varphi_{12}(t, \lambda_*)$ is an eigenfunction for λ_*.)

We will call the function $\Delta(\lambda)$, unique up to a non-zero constant multiple, the characteristic function of the SLP for its importance. The formula in Proposition A.2.1 can be verified by direct calculations using (A.2.3) and will prove to be very useful.

Proposition A.2.1 The characteristic function for the Sturm-Liouville problem consisting of (A.1.1) and (A.1.18) is

$$\Delta(\lambda) = \det \boldsymbol{A} + \det \boldsymbol{B} + \sum_{i,j=1}^{2} c_{ij}\varphi_{ij}(b, \lambda) \tag{A.2.7}$$

where

$$\begin{pmatrix} c_{11} & c_{12} \\ c_{21} & c_{22} \end{pmatrix} = \boldsymbol{B}^T \boldsymbol{A}^C = \begin{pmatrix} a_{22}b_{11} - a_{12}b_{21} & a_{11}b_{21} - a_{21}b_{11} \\ a_{22}b_{12} - a_{12}b_{22} & a_{11}b_{22} - a_{21}b_{12} \end{pmatrix} \tag{A.2.8}$$

with $\boldsymbol{A}^C$ being the matrix of cofactors from $\boldsymbol{A}$. In particular, the characteristic functions for the Dirichlet boundary condition, the Dirichlet-Neumann boundary condition, the Neumann-Dirichlet boundary condition and the Neumann boundary condition are $\varphi_{12}(b, \lambda)$, $\varphi_{22}(b, \lambda)$, $\varphi_{11}(b, \lambda)$ and $\varphi_{21}(b, \lambda)$, respectively.

Next, we prove the entireness of the characteristic function.

Theorem A.2.2 Let $y = y(t, \lambda)$ be a solution to (A.1.1) with $y(a, \lambda)$ and $(py')(a, \lambda)$ independent of λ. Then, on $[a, b] \times \mathbb{C}$

$$\partial_\lambda y = \varphi_{11}\gamma_2 - \varphi_{12}\gamma_1, \quad \partial_\lambda y(py') = \varphi_{21}\gamma_2 - \varphi_{22}\gamma_1 \tag{A.2.9}$$

where

$$\gamma_i(t, \lambda) = \int_a^t \varphi_{1i}(s, \lambda)y(s, \lambda)w(s)\mathrm{d}s, \quad i = 1, 2. \tag{A.2.10}$$

Moreover, for each $t \in [a, b]$, $y(t, \lambda)$ and $py'(t, \lambda)$ are entire functions of λ.

Corollary A.2.1 For each $t \in [a, b]$, $\boldsymbol{\Phi}(t, \lambda)$ is an entire function of λ and

$$\partial_\lambda \boldsymbol{\Phi}(t, \lambda) = \boldsymbol{\Phi}(t, \lambda) \begin{pmatrix} a_{21}(t, \lambda) & a_{22}(t, \lambda) \\ -a_{11}(t, \lambda) & -a_{12}(t, \lambda) \end{pmatrix} \tag{A.2.11}$$

where

$$a_{ij}(t,\lambda) = \int_a^t \varphi_{1i}(s,\lambda)\varphi_{1j}(s,\lambda)w(s)\mathrm{d}s, \quad i,j = 1,2. \quad (A.2.12)$$

In particular, the characteristic function $\Delta(\lambda)$ is entire. Therefore, either all the complex numbers are eigenvalues or the eigenvalues are isolated and do not have an accumulation point in $\mathbb{C}$.

The analytic multiplicity of an isolated eigenvalue is the order of the eigenvalue as a zero of $\Delta(\lambda)$. An eigenvalue is said to be simple if it has multiplicity 1, while the eigenvalues of multiplicity 2 are called double eigenvalues. When we count the (isolated) eigenvalues in a domain in $\mathbb{C}$ of an S-L problem, their multiplicities will be taken into account.

By the above corollary, the reality of $\Phi(b,\lambda)$ for $\lambda \in \mathbb{R}$ when p, q and w are real-valued implies the following result.

Proposition A.2.2 *If p, q and w in (A.1.1) are real-valued, then the non-real eigenvalues for a real boundary condition appear in conjugate pairs and each such pair share the same analytic multiplicity and the same geometric multiplicity.*

Appendix B Thomson-Haskell method

A numerical method is presented here for calculating the Sturm-Liouville problems with piecewise constant coefficients, see [136].

Consider a stratified fluid of finite depth with a rigid top and bottom, but extending infinitely in the horizontal direction. The vertical continuously stratified internal wave equation is

$$\phi''(z) + k_3^2 \phi(z) = 0, \tag{B.1}$$

where

$$k_3^2 = k_h^2 \left(\frac{N^2 - \omega^2}{\omega^2 - f^2} \right). \tag{B.2}$$

In general, the Brunt-Väisälä frequency N will be a continuous function of the depth. The layer thickness and the Brunt-Väisälä frequency vary from layer to layer. In layer p, (B.2) becomes

$$k_{3p}^2 = k_h^2 \left(\frac{N_p^2 - \omega^2}{\omega^2 - f^2} \right), \tag{B.3}$$

and $\phi''(z) + k_{3p}^2 \phi(z) = 0$ has the solution

$$\phi = A_p \exp(m_p z) + B_p \exp(-m_p z), \tag{B.4}$$

where

$$m_p = i k_{3p} \tag{B.5}$$

obviously. In the case where $w > f$, m_p will be real for $w > N_p$ and imaginary for $w < N_p$. Differentiating (B.4) with respect to z, we have

$$\frac{d\phi}{dz} = \phi' = m_p A_p \exp(m_p z) - m_p B_p \exp(-m_p z). \tag{B.6}$$

At the $p - 1$ interface ϕ and ϕ' are

$$\begin{cases} \phi_{p-1} = A_p + B_p, \\ \phi'_{p-1} = m_p A_p - m_p B_p, \end{cases} \tag{B.7}$$

This can be written in the matrix form as

$$\left(\phi_{p-1}, \phi'_{p-1} \right) = \boldsymbol{E}_p \left(A_p, B_p \right), \tag{B.8}$$

or
$$\begin{pmatrix} \phi_{p-1} \\ \phi'_{p-1} \end{pmatrix} = \begin{pmatrix} 1 & 1 \\ m_p & -m_p \end{pmatrix} \begin{pmatrix} A_p \\ B_p \end{pmatrix}, \tag{B.9}$$

where
$$\boldsymbol{E}_p = \begin{pmatrix} 1 & 1 \\ m_p & -m_p \end{pmatrix}. \tag{B.10}$$

At the p interface, $z = h_p$ and
$$\begin{cases} \phi_p = A_p \exp(m_p h_p) + B_p \exp(-m_p h_p) \\ \phi'_p = m_p A_p \exp(m_p h_p) - m_p B_p \exp(-m_p h_p) \end{cases} \tag{B.11}$$

or in the matrix form as
$$(\phi_p, \phi'_p) = \boldsymbol{D}_p (A_p, B_p), \tag{B.12}$$

where
$$\boldsymbol{D}_p = \begin{pmatrix} \exp(m_p h_p) & \exp(-m_p h_p) \\ m_p \exp(m_p h_p) & -m_p \exp(-m_p h_p) \end{pmatrix}. \tag{B.13}$$

Combining
$$(A_p, B_p) = \boldsymbol{E}_p^{-1} (\phi_{p-1}, \phi'_{p-1}),$$
$$(A_p, B_p) = \boldsymbol{D}_p^{-1} (\phi_p, \phi'_p)$$

to eliminate (A_p, B_p), then
$$(\phi_p, \phi'_p) = \boldsymbol{D}_p \boldsymbol{E}_p^{-1} (\phi_{p-1}, \phi'_{p-1}), \tag{B.14}$$

where
$$\boldsymbol{E}_p^{-1} = \begin{pmatrix} \dfrac{1}{2} & \dfrac{1}{2} m_p^{-1} \\ \dfrac{1}{2} & -\dfrac{1}{2} m_p^{-1} \end{pmatrix}. \tag{B.15}$$

A new quantity
$$\boldsymbol{a}_p = \begin{pmatrix} a_{p11} & a_{p12} \\ a_{p21} & a_{p\,12} \end{pmatrix} = \boldsymbol{D}_p \boldsymbol{E}_p^{-1} \tag{B.16}$$

is defined. The terms of the matrix $\boldsymbol{a}_p$ will be hyperbolic sines and cosines when m_p is real, and trigonometric sines and cosines when m_p is imaginary. As an illustration, consider the matrix element $\boldsymbol{a}_p$. For m_p real

Appendix B Thomson-Haskell method

$$a_p = \begin{pmatrix} \operatorname{ch}(m_p h_p) & \dfrac{1}{m_p}\operatorname{sh}(m_p h_p) \\ m_p \operatorname{sh}(m_p h_p) & \operatorname{ch}(m_p h_p) \end{pmatrix}. \tag{B.17}$$

for m_p imaginary

$$a_p = \begin{pmatrix} \cos(k_{3p} h_p) & \dfrac{1}{k_{3p}}\sin(k_3 h_p) \\ -k_{3p} \sin(k_{3p} h_p) & \cos(k_{3p} h_p) \end{pmatrix} \tag{B.18}$$

where

$$k_{3p} = k_{hp}\left(\frac{N_p^2 - \omega^2}{\omega^2 - f^2}\right)^{\frac{1}{2}}. \tag{B.19}$$

Thus, the relationship of the two interfaces bounding layer p is

$$(\phi_p, \phi_p') = a_p (\phi_{p-1}, \phi_{p-1}'). \tag{B.20}$$

In a similar manner for the $p-1$ layer of thickness ϕ_{p-1} and frequency N_{p-1} and bounded by the $p-2$ and $p-1$ interfaces, is obtained

$$(\phi_{p-1}, \phi_{p-1}') = a_{p-1}(\phi_{p-2}, \phi_{p-2}'). \tag{B.21}$$

Because (as imposed conditions) both ϕ and ϕ' are required to be continuous across interfaces between layers, the values of ϕ_{p-1}, ϕ_{p-1}' are the same for (B.20) and (B.21) and the layer to which they refer does not have to be denoted as long as their values at the interface only are used.

Equations (B.20) and (B.21) can be combined to obtain

$$(\phi_p, \phi_p') = a_p a_{p-1}(\phi_{p-2}, \phi_{p-2}'). \tag{B.22}$$

If this process is continued upward to the first layer, bounded by the 0 and 1 interfaces and downward to the last n layer, bounded by the $n-1$ and n interfaces, the resulting equation is

$$(\phi_n, \phi_n') = a_n a_{n-1} \cdots a_{p+1} a_p a_{p-1} \cdots a_2 a_1 (\phi_0, \phi_0')$$
$$= F(\phi_0, \phi_0'), \tag{B.23}$$

where

$$F = \begin{pmatrix} F_{11} & F_{12} \\ F_{21} & F_{22} \end{pmatrix} \tag{B.24}$$

$$= a_n a_{n-1} \cdots a_2 a_1.$$

Here ϕ_n, ϕ_0 are the values of ϕ at the bottom and surface, respectively; ϕ_n is set equal to zero because of the rigid bottom, and ϕ_0 is assumed to be zero because surface motions of internal waves are very small. Thus,

$$\phi_0 = \phi_n = 0. \tag{B.25}$$

Equation (B.23) can be written as

$$0 = 0F_{11} + F_{12}\phi_0' \tag{B.26}$$

and

$$\phi_n' = 0F_{21} + F_{22}\phi_0'. \tag{B.27}$$

Either of the two equations can be used as a period equation. Equation (B.26) is chosen because it yields the simpler one

$$F_{12} = 0. \tag{B.28}$$

Equation (B.28) can be solved numerically using an iteration technique.

For a given ω and mode number when the value of m_p is real, ϕ and ϕ' in layer p can be written in terms of exponentials

$$\phi_p = A_p \exp(m_p z) + B_p \exp(-m_p z),$$

$$\phi_p' = m_p A_p \exp(m_p z) - m_p B_p \exp(-m_p z).$$

For m_p imaginary, it is more convenient to employ trigonometric functions

$$\phi_p = A_p \sin(r_p z) + B_p \cos(r_p z),$$

$$\phi_p' = r_p A_p \cos(r_p z) - r_p B_p \sin(r_p z),$$

where $r_p = -im_p$ real.

At the bottom in layer n, $z = z_n$, $\phi_n = 0$. With m_n real

$$B = -A_n \exp(2m_n z_n). \tag{B.29}$$

For m_n imaginary, $r_n = im_n$ is real and

$$B = -A_n \tan(r_n z_n). \tag{B.30}$$

A_n must be chosen at the start of the calculation and all remaining constants (A_p, B_p) will be in terms of it. At any interface, both ϕ and ϕ' are continuous. It is easily seen that all the constants can be determined in terms of A_n. Once the constants are determined, the values of ϕ and ϕ' at any depth can be found.

Appendix C First-order linear differential equations

This appendix originated from related sections in [3].

Throughout this appendix, $J \subseteq \mathbb{R}$ will always be an interval (open or half-open or closed, bounded or unbounded) with left endpoint a and right endpoint b, where

$$-\infty \leqslant a < b \leqslant +\infty.$$

We use $\mathrm{AC}_{\mathrm{loc}}(J,\mathbb{C})$ to denote the set of complex-valued functions which are absolutely continuous in all compact subintervals of J, and $\mathrm{AC}_{\mathrm{loc}}(J,\mathbb{R})$ is defined in a similar way.

It is also always assumed that $n, m, l \in \mathbb{N}$. We use $M_{n,m}^{\mathbb{C}}$ to denote the vector space of n by m matrices with complex entries, and $M_{n,m}^{\mathbb{R}}$ has a similar meaning. Then, the definitions of $L(J, M_{n,m}^{\mathbb{C}})$, $L(J, M_{n,m}^{\mathbb{R}})$, $L_{\mathrm{loc}}(J, M_{n,m}^{\mathbb{C}})$, $L_{\mathrm{loc}}(J, M_{n,m}^{\mathbb{R}})$, $\mathrm{AC}_{\mathrm{loc}}(J, M_{n,m}^{\mathbb{C}})$ and $\mathrm{AC}_{\mathrm{loc}}(J, M_{n,m}^{\mathbb{R}})$ are clear.

For each $\boldsymbol{P} = (p_{i,j}) : J \to M_{n,m}^{\mathbb{C}}$, we define

$$|\boldsymbol{P}| = \sum |p_{i,j}|.$$

Then, for any $\boldsymbol{P} : J \to M_{n,m}^{\mathbb{C}}$ and $\boldsymbol{Q} : J \to M_{m,l}^{\mathbb{C}}$,

$$|\boldsymbol{PQ}| \leqslant |\boldsymbol{P}||\boldsymbol{Q}|.$$

C.1 Existence and uniqueness of a solution

Given $\boldsymbol{P} : J \to M_{n,n}^{\mathbb{C}}$ and $\boldsymbol{F} : J \to M_{n,m}^{\mathbb{C}}$, consider the differential equation

$$\boldsymbol{Y}' = \boldsymbol{PY} + \boldsymbol{F} \quad \text{on } J, \tag{C.1.1}$$

By a solution of (C.1.1) we mean a function $\boldsymbol{Y} \in \mathrm{AC}_{\mathrm{loc}}(J, M_{n,m}^{\mathbb{C}})$ satisfying (C.1.1) a.e.. We now show the existence and uniqueness of a solution to any initial value problem (IVP) under the local integrability of the coefficient functions.

Theorem C.1.1 If

$$\boldsymbol{P} \in L_{\mathrm{loc}}(J, M_{n,n}^{\mathbb{C}}), \qquad \boldsymbol{F} \in L_{\mathrm{loc}}(J, M_{n,m}^{\mathbb{C}}), \tag{C.1.2}$$

then for any $t_0 \in J$ and any $C \in M_{n,m}^{\mathbb{C}}$, the IVP

$$Y' = PY + F, \quad Y(t_0) = C \qquad (C.1.3)$$

has a unique solution Y on the whole J, and the solution Y satisfies

$$|Y(t) - C| \leq \left|\int_{t_0}^{t} |PC + F|\right| e^{\left|\int_{t_0}^{t} |P|\right|}, \quad t \in J. \qquad (C.1.4)$$

Moreover, if C, P and F are real-valued, so is the solution Y.

Proof. Note that if Y is a solution to the IVP in an interval $J_0 \subseteq J$ containing t_0, then

$$Y(t) = C + \int_{t_0}^{t} (PY + F), \quad t \in J_0. \qquad (C.1.5)$$

Conversely, every continuous function Y in some interval $J_0 \subseteq J$ containing t_0 satisfying (C.1.5) is also a solution to the IVP on J_0, since (C.1.5) then implies that $Y \in \mathrm{AC}_{\mathrm{loc}}(J_0, M_{n,m}^{\mathbb{C}})$.

To construct a solution to the IVP by successive approximations, we define for $t \in J$,

$$Y_0(t) = C, \qquad (C.1.6)$$

$$Y_i(t) = C + \int_{t_0}^{t} (PY_{i-1} + F), \quad i \in \mathbb{N}. \qquad (C.1.7)$$

Then, for each $i \in \mathbb{N}$, Y_i is a continuous function on J and

$$Y_{i+1}(t) - Y_i(t) = \int_{t_0}^{t} P(Y_i - Y_{i-1}), \quad t \in J. \qquad (C.1.8)$$

Assume that t_0 is not the right endpoint of J and let $t_1 \in J$ satisfying $t_1 > t_0$. For $t \in (t_0, t_1)$, define

$$\sigma(t) = \int_{t_0}^{t} |P|, \quad \tau(t) = \int_{t_0}^{t} |PC + F|. \qquad (C.1.9)$$

Since σ and τ are non-decreasing on $[t_0, t_1]$, we have that for $t \in [t_0, t_1]$,

$$|Y_1(t) - Y_0(t)| = \left|\int_{t_0}^{t} (PC + F)\right| \leq \tau(t), \qquad (C.1.10)$$

$$|Y_2(t) - Y_1(t)| = \left|\int_{t_0}^{t} P(Y_1 - Y_0)\right| \leq \int_{t_0}^{t} |P|\tau$$

$$\leq \tau(t) \int_{t_0}^{t} |P| = \sigma(t)\tau(t), \qquad (C.1.11)$$

$$|Y_3(t) - Y_2(t)| = \left|\int_{t_0}^t P(Y_2 - Y_1)\right| \leqslant \int_{t_0}^t |P|\sigma\tau$$

$$\leqslant \tau(t)\int_{t_0}^t |P|\sigma = \frac{\sigma(t)^2}{2!}\tau(t), \tag{C.1.12}$$

and in general,

$$|Y_{i+1}(t) - Y_i(t)| = \frac{\sigma(t)^i}{i!}\tau(t) \leqslant \frac{\sigma(t_1)^i}{i!}\tau(t_1), \quad i \in \mathbb{N}_0. \tag{C.1.13}$$

Since

$$\sum_{i=0}^{+\infty} \frac{\sigma(t_1)^i}{i!} = e^{\sigma(t_1)}, \tag{C.1.14}$$

from

$$|Y_{i+j}(t) - Y_i(t)| \leqslant \tau(t_1)\sum_{k=i}^{i+j-1} \frac{\sigma(t_1)^k}{k!}, \quad i \in \mathbb{N}_0, j \in \mathbb{N}, t \in [t_0, t_1], \tag{C.1.15}$$

we see that the sequence Y_i of continuous functions converges uniformly on $[t_0, t_1]$, say to Y. Then, the continuous function Y satisfies (C.1.5) on $[t_0, t_1]$ by (C.1.7) and

$$|Y(t_1) - C| \leqslant \tau(t_1)e^{\sigma(t_1)} \tag{C.1.16}$$

by (C.1.15). To show that this Y is the unique solution to the IVP on $[t_0, t_1]$, let Z be any continuous function satisfying (C.1.15) on $[t_0, t_1]$. Then, $|Y - Z|$ is bounded on $[t_0, t_1]$, say by $\delta > 0$. For any $t \in [t_0, t_1]$,

$$|Y(t) - Z(t)| = \left|\int_{t_0}^t P(Y - Z)\right| \leqslant \delta\int_{t_0}^t |P| = \sigma(t)\delta, \tag{C.1.17}$$

and proceeding as above we get

$$|Y(t) - Z(t)| \leqslant \frac{\sigma(t_1)^i}{i!}\delta, \quad \forall i \in \mathbb{N}, \tag{C.1.18}$$

which implies that $Y(t) = Z(t)$. Since $t_1 > t_0$ is arbitrary, this shows the existence and uniqueness of a solution together with the inequality (C.1.4) on the part of J right to t_0. The proof for the part of J left to t_0 is similar.

The moreover statement is clear. □

Remark C.1.1 The proof of Theorem C.1.1 actually implies the following.

(1) Consider the case where the left endpoint a of J is finite. Assume, in addition to (C.1.2), that

$$P \in L\big((a, b_0), M_{n,n}^{\mathbb{C}}\big), \quad F \in L\big((a, b_0), M_{n,m}^{\mathbb{C}}\big), \tag{C.1.19}$$

for some (and hence all) $b_0 \in J$. By replacing J by the interval $\{a\} \cup J$ if $a \notin J$, we see: first, any solution to (C.1.1) can be extended to a by continuity to become absolutely continuous on $[a, b_0]$ for all $b_0 \in J$, which is the definition of $\boldsymbol{Y}(a)$ that we will take from now on; second, for any $t_0 \in \{a\} \cup J$ and any $\boldsymbol{C} \in M_{n,m}^{\mathbb{C}}$, there is a unique solution $\boldsymbol{Y}$ to (C.1.1) such that $\boldsymbol{Y}(t_0) = \boldsymbol{C}$.

(2) We have similar results for the right endpoint b of J and a similar definition of $\boldsymbol{Y}(b)$ when b is finite.

In particular, if a, b are both finite and

$$\boldsymbol{P} \in L\big((a,b), M_{n,n}^{\mathbb{C}}\big), \qquad \boldsymbol{F} \in L\big((a,b), M_{n,m}^{\mathbb{C}}\big), \tag{C.1.20}$$

then: first, any solution to (C.1.1) can be extended to both a and b by continuity to become absolutely continuous on the interval $[a, b]$; second, for any $t_0 \in [a, b]$ and any $\boldsymbol{C} \in M_{n,m}^{\mathbb{C}}$, there is a unique solution $\boldsymbol{Y}$ to (C.1.1) such that $\boldsymbol{Y}(t_0) = \boldsymbol{C}$.

Motivated by Remark C.1.1, we make the following definitions.

Definition C.1.1 The left endpoint a (finite or infinite) of the differential equation (C.1.1) is said to be regular if, in addition to (C.1.2),

$$\boldsymbol{P} \in L\big((a,b_0), M_{n,n}^{\mathbb{C}}\big), \qquad \boldsymbol{F} \in L\big((a,b_0), M_{n,m}^{\mathbb{C}}\big), \tag{C.1.21}$$

for some (and hence all) $b_0 \in J$; otherwise, a is called singular. Similar definitions are made for the right endpoint b. The differential equation (C.1.1) is said to be regular if both of its endpoints are regular, i.e., if

$$\boldsymbol{P} \in L(J, M_{n,n}^{\mathbb{C}}), \qquad \boldsymbol{F} \in L(J, M_{n,m}^{\mathbb{C}}). \tag{C.1.22}$$

To generalize the results in Remark of Theorem C.1.1 to the case where the endpoint in question is infinite, we consider the substitution

$$t = t(s) = \int_{-\infty}^{s} f(r)\mathrm{d}r, \tag{C.1.23}$$

where $f \in L^{+}(\mathbb{R}, \mathbb{R}) =: \{f \in L(\mathbb{R}, \mathbb{R}); f > 0 \text{ a.e. on } \mathbb{R}\}$. After the substitution, the differential equation (C.1.1) is transformed to the differential equation

$$\frac{\mathrm{d}}{\mathrm{d}s}\boldsymbol{Y}\big(t(s)\big) = f(s)\boldsymbol{P}\big(t(s)\big)\boldsymbol{Y}\big(t(s)\big) + f(s)\boldsymbol{F}\big(t(s)\big) \quad \text{on } I, \tag{C.1.24}$$

where I is the interval for the variable s corresponding to the interval J for the variable t, i.e., $I = (c, d)$ if $J = (a, b)$, $I = (c, d]$ if $J = (a, b]$, etc. with

$$c = \int_{-\infty}^{a} f(s)\mathrm{d}s \quad \text{and} \quad d = \int_{-\infty}^{b} f(s)\mathrm{d}s \tag{C.1.25}$$

being finite. A function $\boldsymbol{Y}(\cdot)$ is a solution to (C.1.1) if and only if $\boldsymbol{Y}(t(\cdot))$ is a solution to (C.1.24). Note that (C.1.2) holds if and only if

$$f(\cdot)\boldsymbol{P}(t(\cdot)) \in L_{\mathrm{loc}}(I, M_{n,n}^{\mathbb{C}}), \qquad f(\cdot)\boldsymbol{F}(t(\cdot)) \in L_{\mathrm{loc}}(I, M_{n,m}^{\mathbb{C}}). \tag{C.1.26}$$

Similarly, (C.1.1) is regular at a (b, respectively) if and only if (C.1.24) is regular at c (d, respectively). Thus, the above substitution implies that the results of Remark C.1.1 of Theorem C.1.1 are still true even when the endpoint in question is infinite, if we generalize the definition of absolutely continuous functions by

$$\mathrm{AC}\left((a,b), M_{n,m}^{\mathbb{C}}\right) = \mathrm{AC}_{\mathrm{loc}}\left((a,b), M_{n,m}^{\mathbb{C}}\right), \tag{C.1.27}$$

$$\mathrm{AC}\left([a,b), M_{n,m}^{\mathbb{C}}\right) = \Big\{\boldsymbol{Y}: [a,b) \to M_{n,m}^{\mathbb{C}};\ \boldsymbol{Y}' \in L\left((a,t), M_{n,m}^{\mathbb{C}}\right)$$
$$\boldsymbol{Y}(t) = \boldsymbol{Y}(a) + \int_a^t \boldsymbol{Y}',\ \forall t \in (a,b)\Big\}, \tag{C.1.28}$$

$$\mathrm{AC}((a,b], M_{n,m}^{\mathbb{C}}) = \Big\{\boldsymbol{Y}: (a,b] \to M_{n,m}^{\mathbb{C}};\ \boldsymbol{Y}' \in L\left((t,b), M_{n,m}^{\mathbb{C}}\right)$$
$$\boldsymbol{Y}(t) = \boldsymbol{Y}(b) - \int_t^b \boldsymbol{Y}',\ \forall t \in (a,b)\Big\}, \tag{C.1.29}$$

$$\mathrm{AC}([a,b], M_{n,m}^{\mathbb{C}}) = \Big\{\boldsymbol{Y}: [a,b] \to M_{n,m}^{\mathbb{C}};\ \boldsymbol{Y}' \in L\left((a,b), M_{n,m}^{\mathbb{C}}\right)$$
$$\boldsymbol{Y}(t) = \boldsymbol{Y}(a) + \int_a^t \boldsymbol{Y}',\ \forall t \in (a,b)\Big\}. \tag{C.1.30}$$

Note that here we have extended the concept of the interval to include $[\infty, b)$, etc. (however, for consistency, the notation J will not be used for any of these new intervals). So we have the following.

Theorem C.1.2 Assume that (C.1.2) holds.

(1) If (C.1.1) is regular at its left endpoint a, then: first, any solution to (C.1.1) can be extended to a by continuity to become absolutely continuous in the interval $\{a\} \cup J$, which is the definition of $\boldsymbol{Y}(a)$ that we will take from now on; second, for any $t_0 \in \{a\} \cup J$ and any $\boldsymbol{C} \in M_{n,m}^{\mathbb{C}}$, there is a unique solution $\boldsymbol{Y}$ to (C.1.1) such that $\boldsymbol{Y}(t_0) = \boldsymbol{C}$.

(2) We have similar results for the right endpoint b of (C.1.1) and a similar definition of $\boldsymbol{Y}(b)$.

In particular, if (C.1.1) is regular, then: first, any solution to (C.1.1) can be extended to both a and b by continuity to become absolutely continuous on the interval $[a,b]$; second, for any $t_0 \in [a,b]$ and any $\boldsymbol{C} \in M_{n,m}^{\mathbb{C}}$, there is a unique solution $\boldsymbol{Y}$ to (C.1.1) such that $\boldsymbol{Y}(t_0) = \boldsymbol{C}$.

C.2 Rank of a solution and variation of parameters

In this section, we discuss the rank of a solution, introduce the concept of fundamental solutions, and present the so-called variation of parameter formula.

Theorem C.2.1 Assume that
$$P \in L_{\text{loc}}(J, M_{n,n}^{\mathbb{C}}) \tag{C.2.1}$$
and $Y: J \to M_{n,m}^{\mathbb{C}}$ is a solution to
$$Y' = PY \quad \text{on } J. \tag{C.2.2}$$
Then, rank Y is constant on J. Moreover, if $n = m$ and $t_0 \in J$, then
$$\det Y(t) = (\det Y(t_0)) \exp\left(\int_{t_0}^t \text{tr} P\right), \quad t \in J. \tag{C.2.3}$$

Proof. When $n = m$, from (C.2.2) we obtain that
$$(\det Y)' = (\text{tr} P) \det Y \quad \text{a.e. on } J \tag{C.2.4}$$
which implies (C.2.3).

To prove the rank claim, let $t_0 \in J$ and set $r = \text{rank } Y(t_0)$. If $r = 0$, then $Y \equiv 0$ on J, so rank $Y \equiv 0$ on J. If $r > 0$, let $C_1, C_2, \cdots, C_r$ be linearly independent columns of $Y(t_0)$ and construct a non-singular $n \times n$ matrix D by putting $n \times r$ appropriate columns next to $C_1, C_2, \cdots, C_r$ if $r < n$. Denote by Z the solution to $Z' = PZ$ on J satisfying $Z(t_0) = D$. Then, rank $Z \equiv n$ on J. The first r columns of Z are just the r columns of Y corresponding to $C_1, C_2, \cdots, C_r$, respectively. Since these columns are linearly independent, we have rank $Y \geqslant r$ on J. This actually implies that rank $Y \equiv r$ on J: if rank $Y(t_1) > r$ for some $t_1 \in J$, then rank $Y \geqslant \text{rank } Y(t_1) > r$ on J, which is impossible by the choice of r. □

Combining Theorem C.1.2 and Theorem C.2.1 we have the following.

Theorem C.2.2 Let the assumptions of Theorem C.2.1 hold.

(1) If (C.2.2) is regular at its left endpoint a, then J in the conclusions of Theorem C.2.1 can be replaced by the interval $\{a\} \cup J$.

(2) If (C.2.2) is regular at its right endpoint b, then J in the conclusions of Theorem C.2.1 can be replaced by the interval $J \cup \{b\}$.

In particular, if (C.2.2) is regular, then J in the conclusions of Theorem C.2.1 can be replaced by the interval $[a, b]$.

The following two theorems of Everitt and Race show that in order for every IVP to have a solution, the coefficient functions of a linear differential equation must be locally integrable.

Theorem C.2.3 Let $P: J \to M_{n,n}^{\mathbb{C}}$. Assume that there are $t_0 \in J$ and linearly independent vectors $C_1, C_2, \cdots, C_n \in \mathbb{C}^n$ such that each initial value problem

$$Y' = PY, \quad Y(t_0) = C_i \tag{C.2.5}$$

has a solution Y_i on J. Then,

$$P \in L_{\text{loc}}(J, M_{n,n}^{\mathbb{C}}). \tag{C.2.6}$$

Moreover, if each Y_i is C^1, then P is continuous (after its values on a measure 0 subset of J are changed when necessary).

Proof. Let Y be the matrix whose i-th column is Y_i for $i = 1, 2, \cdots, n$. Then, we have (C.2.4), which implies that $\text{tr} P \in L_{\text{loc}}(J, \mathbb{C})$ and

$$\det Y(t) = (\det Y(t_0)) \exp\left(\int_{t_0}^t \text{tr} P\right) \neq 0, \quad t \in J. \tag{C.2.7}$$

Since for each compact interval $J_0 \subseteq J$ containing t_0 on which $\det Y$ is never 0, the continuous function $1/\det Y$ is bounded, while $(\det Y)'$ is locally Lebsegue integrable on J by the definition of solutions. Thus,

$$P = Y'Y^{-1} \in L_{\text{loc}}(J, M_{n,n}^{\mathbb{C}}), \tag{C.2.8}$$

which also implies the moreover statement. □

Theorem C.2.4 Let $P: J \to M_{n,n}^{\mathbb{C}}$ and $F: J \to \mathbb{C}^n$. Assume that there are $t_0 \in J$ and linearly independent vectors $C_1, C_2, \cdots, C_n \in \mathbb{C}^n$ such that for each $i \in \{0, 1, \cdots, n\}$, the initial value problem

$$Y' = PY + F, \quad Y(t_0) = C_i \tag{C.2.9}$$

has a solution Y_i on J, where $C_0 = 0$. Then,

$$P \in L_{\text{loc}}(J, M_{n,n}^{\mathbb{C}}), \quad F \in L_{\text{loc}}(J, \mathbb{C}^n). \tag{C.2.10}$$

Moreover, if each Y_i is C^1, then P and F are continuous (after their values on a measure 0 subset of J are changed when necessary).

Proof. Note that for $i = 1, 2, \cdots, n$, $Y_i - Y_0$ is the solution to (C.2.2) satisfying $Y(t_0) = C_i$. Thus, (C.2.6) holds by Theorem C.2.3. So

$$F = Y_0' - PY_0 \in L_{\text{loc}}(J, \mathbb{C}^n), \tag{C.2.11}$$

which together with Theorem C.2.3 also implies the moreover statement. □

Proposition C.2.1 Assume that (C.2.6) holds. If $\boldsymbol{\Phi}\colon J \to M_{n,n}^{\mathbb{C}}$ is a solution to (C.2.2) such that $\boldsymbol{\Phi}(t)$ is non-singular for some (and hence all) $t \in J$, then for any $t_0 \in J$ and any $\boldsymbol{C} \in M_{n,m}^{\mathbb{C}}$, $\boldsymbol{Y} = \boldsymbol{\Phi}\boldsymbol{\Phi}(t_0)^{-1}\boldsymbol{C}$, is the solution to (C.2.2) satisfying $\boldsymbol{Y}(t_0) = \boldsymbol{C}$.

Proof. Direct substitution shows that $\boldsymbol{Y}$ is a solution to (C.2.2). Clearly, $\boldsymbol{Y}(t_0) = \boldsymbol{C}$. □

Corollary C.2.1 Assume that (C.2.6) holds. Then, all the solutions to (C.2.2) form an n-dimensional vector space over $\mathbb{C}$. Moreover, if $\boldsymbol{P}$ is real-valued, then all the real solutions to (C.2.2) form an n-dimensional vector space over $\mathbb{R}$.

Proof. The columns $\boldsymbol{\Phi}_1, \boldsymbol{\Phi}_2, \cdots, \boldsymbol{\Phi}_n$ of any $\boldsymbol{\Phi}$ used in Proposition C.2.1 are n linearly independent such solutions, and Proposition C.2.1 shows that any other such solution is a linear combination of $\boldsymbol{\Phi}_1, \boldsymbol{\Phi}_2, \cdots, \boldsymbol{\Phi}_n$. □

Remark C.2.1 If (C.2.2) is regular at a, then t_0 in Proposition C.2.1 can also be a. A similar result for b is true. In particular, if (C.2.2) is regular, then t_0 can also be any of a and b.

Definition C.2.1 Any solution $\boldsymbol{Y}\colon J \to M_{n,n}^{\mathbb{C}}$ of (C.2.2) that is non-singular at some (and hence any) point of J is called a fundamental solution to (C.2.2).

To end this section, we prove the variation of parameter formula, which shows that when a fundamental solution to the homogeneous equation (C.2.2) is known, one can even obtain the solution to any IVP of any associated nonhomogeneous equation.

Theorem C.2.5 Assume that

$$\boldsymbol{P} \in L_{\text{loc}}(J, M_{n,n}^{\mathbb{C}}), \qquad \boldsymbol{F} \in L_{\text{loc}}(J, M_{n,m}^{\mathbb{C}}). \tag{C.2.12}$$

and $\boldsymbol{\Phi}\colon J \to M_{n,n}^{\mathbb{C}}$ is a fundamental solution to (C.2.2). Then, for any $t_0 \in J$ and any $\boldsymbol{C} \in M_{n,m}^{\mathbb{C}}$,

$$\boldsymbol{Y}(t) = \boldsymbol{\Phi}(t)\boldsymbol{\Phi}(t_0)^{-1}\boldsymbol{C} + \boldsymbol{\Phi}(t)\int_{t_0}^{t} \boldsymbol{\Phi}(s)^{-1}\boldsymbol{F}(s)\mathrm{d}s, \quad t \in J \tag{C.2.13}$$

is the solution to

$$\boldsymbol{Y}' = \boldsymbol{PY} + \boldsymbol{F} \quad \text{on } J \tag{C.2.14}$$

satisfying $\boldsymbol{Y}(t_0) = \boldsymbol{C}$.

Proof. From (C.2.13) and the choice of $\boldsymbol{\Phi}$ we obtain that for almost all $t \in J$,

$$\boldsymbol{Y}'(t) = \boldsymbol{P}(t)\boldsymbol{\Phi}(t)\boldsymbol{\Phi}(t_0)^{-1}\boldsymbol{C} + \boldsymbol{P}(t)\boldsymbol{\Phi}(t)\int_{t_0}^{t} \boldsymbol{\Phi}(s)^{-1}\boldsymbol{F}(s)\mathrm{d}s + \boldsymbol{F}(t)$$

$$= \boldsymbol{P}(t)\boldsymbol{Y}(t) + \boldsymbol{F}(t), \tag{C.2.15}$$

i.e., $\boldsymbol{Y}$ is a solution to (C.2.14). Clearly, $\boldsymbol{Y}(t_0) = \boldsymbol{C}$. □

Remark C.2.2 If (C.2.14) is regular at a, then t_0 in Theorem C.2.5 can also be a. A similar result for b is true. In particular, if (C.2.14) is regular, then t_0 can also be any of a and b.

C.3 Continuous dependence of solution on the problem

In this section, we show that the solution of an IVP depends continuously on the problem, i.e., on t_0, $\boldsymbol{C}$, $\boldsymbol{P}$ and $\boldsymbol{F}$ in the problem, in some appropriate sense.

Proposition C.3.1 Let

$$\boldsymbol{P}, \boldsymbol{Q} \in L_{\mathrm{loc}}(J, M_{n,n}^{\mathbb{C}}), \qquad \boldsymbol{F}, \boldsymbol{G} \in L_{\mathrm{loc}}(J, M_{n,m}^{\mathbb{C}}). \tag{C.3.1}$$

$t_0, t_1 \in J$ and $\boldsymbol{C}, \boldsymbol{D} \in M_{n,m}^{\mathbb{C}}$. If $\boldsymbol{Y}$ and $\boldsymbol{Z}$ are the solutions to

$$\boldsymbol{Y}' = \boldsymbol{P}\boldsymbol{Y} + \boldsymbol{F} \quad \text{on } J \qquad \text{and} \qquad \boldsymbol{Z}' = \boldsymbol{Q}\boldsymbol{Z} + \boldsymbol{G} \quad \text{on } J \tag{C.3.2}$$

satisfying $\boldsymbol{Y}(t_0) = \boldsymbol{C}$ and $\boldsymbol{Z}(t_1) = \boldsymbol{D}$, respectively, then

$$|\boldsymbol{Y}(t) - \boldsymbol{Z}(t) - \boldsymbol{C} + \boldsymbol{D}| \leqslant \left|\int_{t_0}^{t} \left(|\boldsymbol{P}| \cdot |\boldsymbol{C} - \boldsymbol{D}| + |\boldsymbol{P} - \boldsymbol{Q}|\gamma + |\boldsymbol{F} - \boldsymbol{G}|\right)\right| \mathrm{e}^{\left|\int_{t_0}^{t} |\boldsymbol{P}|\right|} +$$

$$\left(\left|\int_{t_0}^{t} |\boldsymbol{P}|\right| \mathrm{e}^{\left|\int_{t_0}^{t} |\boldsymbol{P}|\right|} + 1\right) \left|\int_{t_0}^{t_1} |\boldsymbol{Q}\boldsymbol{D} + \boldsymbol{G}|\right| \mathrm{e}^{\left|\int_{t_0}^{t} |\boldsymbol{Q}|\right|}, \quad t \in J, \tag{C.3.3}$$

where

$$\gamma(s) = |\boldsymbol{D}| + \left|\int_{t_1}^{s} |\boldsymbol{Q}\boldsymbol{D} + \boldsymbol{G}|\right| \mathrm{e}^{\left|\int_{t_1}^{s} |\boldsymbol{Q}|\right|}. \tag{C.3.4}$$

Proof. From (C.3.2) we obtain that

$$(\boldsymbol{Y} - \boldsymbol{Z})' = \boldsymbol{P}(\boldsymbol{Y} - \boldsymbol{Z}) + \left[(\boldsymbol{P} - \boldsymbol{Q})\boldsymbol{Z} + \boldsymbol{F} - \boldsymbol{G}\right] \quad \text{a.e. on } J, \tag{C.3.5}$$

which together with (C.1.4) yields

$$|Y(t) - Z(t) - C + D|$$
$$\leqslant |Y(t) - Z(t) - C + Z(t_0)| + |Z(t_0) - D|$$
$$\leqslant \left| \int_{t_0}^{t} |P(C - Z(t_0)) + (P - Q)Z + F - G| \, e^{\left| \int_{t_0}^{t} |P| \right|} \right| + |Z(t_0) - D|$$
$$\leqslant \left| \int_{t_0}^{t} (|P| |C - D| + |P - Q| |Z| + |F - G|) \, e^{\left| \int_{t_0}^{t} |P| \right|} \right| +$$
$$\left(\left| \int_{t_0}^{t} |P| \right| \, |e^{\left| \int_{t_0}^{t} |P| \right|} + 1 \right) |Z(t_0) - D|, \quad t \in J.$$
(C.3.6)

On the other hand, (C.1.4) also implies that

$$|Z(s)| \leqslant |D| + |Z(s) - D| \leqslant \gamma(s), \tag{C.3.7}$$

$$|Z(t_0) - D| \leqslant \left| \int_{t_0}^{t_1} |QD + G| \right| e^{\left| \int_{t_0}^{t_1} |Q| \right|}. \tag{C.3.8}$$

Combining (C.3.6), (C.3.7) and (C.3.8) one gets (C.3.3). □

Let $\{F_i\}_{i \in \mathbb{N}_0}$ be a sequence in $L_{\text{loc}}(J, M_{n,m}^{\mathbb{C}})$. We say that $F_i \to F_0$ in $L_{\text{loc}}(J, M_{n,m}^{\mathbb{C}})$ as $i \to +\infty$ if for each compact interval $J_0 \subseteq J$ we have

$$\int_{J_0} |F_i - F_0| \to 0 \quad \text{as } i \to +\infty. \tag{C.3.9}$$

The following result is a direct consequence of Proposition C.3.1.

Theorem C.3.1 Let

$$\{P_i\}_{i \in \mathbb{N}_0} \subset L_{\text{loc}}(J, M_{n,n}^{\mathbb{C}}), \quad \{F_i\}_{i \in \mathbb{N}_0} \subset L_{\text{loc}}(J, M_{n,m}^{\mathbb{C}}), \tag{C.3.10}$$

$\{t_i\}_{i \in \mathbb{N}_0} \subset J$ and $\{C_i\}_{i \in \mathbb{N}_0} \subset M_{n,m}^{\mathbb{C}}$. Assume that for $i \in \mathbb{N}_0$, Y_i is the solution to

$$Y_i' = P_i Y_i + F_i \quad \text{on } J \tag{C.3.11}$$

satisfying $Y_i(t_i) = C_i$. If

$$P_i \to P_0 \text{ in } L_{\text{loc}}(J, M_{n,n}^{\mathbb{C}}), \quad F_i \to F_0 \text{ in } L_{\text{loc}}(J, M_{n,m}^{\mathbb{C}}), \quad t_i \to t_0, \quad C_i \to C_0 \tag{C.3.12}$$

as $i \to +\infty$, then

$$Y_i \to Y_0 \quad \text{as } i \to +\infty \tag{C.3.13}$$

locally uniformly on J, i.e., uniformly in each compact subinterval of J.

To compare solutions of different regular differential equations (possibly on different intervals) globally, we extend a regular differential equation

$$\boldsymbol{Y}' = \boldsymbol{PY} + \boldsymbol{F} \quad \text{on } J \tag{C.3.14}$$

to the regular differential equation

$$\boldsymbol{Z}' = \widetilde{\boldsymbol{P}}\boldsymbol{Z} + \widetilde{\boldsymbol{F}} \quad \text{on } \mathbb{R}, \tag{C.3.15}$$

where $\widetilde{\boldsymbol{P}}$ is the extension of $\boldsymbol{P}$ to $\mathbb{R}$ that is equal to 0 on the quotient space $\mathbb{R} \setminus J$, and $\widetilde{\boldsymbol{F}}$ has a similar meaning. Each solution to (C.3.14) can be extended to become a solution to (C.3.15) using its endpoint values. And every solution to (C.3.15) can be obtained in this way. After identifying (C.3.14) with (C.3.15), we see that the space of regular differential equations in dimension $n \times m$ is just

$$L(\mathbb{R}, M_{n,n}^{\mathbb{C}}) \times L(\mathbb{R}, M_{n,m}^{\mathbb{C}}) \tag{C.3.16}$$

a Banach space. With the obvious topology on $[-\infty, +\infty]$, the following theorem is also a direct consequence of Proposition C.3.1.

Theorem C.3.2 As a continuous function on $[-\infty, +\infty]$, the solution $\boldsymbol{Y}$ of the regular differential equation $\boldsymbol{Y}' = \boldsymbol{PY} + \boldsymbol{F}$ on $\mathbb{R}$ satisfying $\boldsymbol{Y}(t_0) = \boldsymbol{C}$ depends continuously on

$$(t_0, \boldsymbol{C}, \boldsymbol{P}, \boldsymbol{F}) \in [-\infty, +\infty] \times M_{n,m}^{\mathbb{C}} \times L(\mathbb{R}, M_{n,n}^{\mathbb{C}}) \times L(\mathbb{R}, M_{n,m}^{\mathbb{C}}). \tag{C.3.17}$$

Remark C.3.1 If we only consider regular differential equations in a given dimension in a fixed interval, there is a similar result.

References

[1] 曹之江. 常微分算子 [M]. 北京：科学出版社, 2017.
CAO Z J. Ordinary Differential Operators[M]. Beijing: Science Press, 2017.

[2] 傅守忠, 王忠, 魏广生. Sturm-Liouville 问题及其逆问题 [M]. 北京：科学出版社, 2015.
FU S Z, WANG Z, WEI G S. Sturm-Liouville Problems and Its Inverse Problems[M]. Beijing: Science Press, 2015.

[3] FU S Z, WANG Z, WU H Y. The Geometric Aspects of Sturm-Liouville Problems[M]. Beijing: Science Press, 2019.

[4] 王桂霞. Sturm-Liouville 问题的谱分析与数值计算 [D]. 呼和浩特: 内蒙古大学, 2008.
WANG G X. Spectral analysis and numerical computation of Sturm-Liouville problems[D]. Huhhot: Inner Mongolia University, 2008.

[5] WANG G X, WANG Z, WU H Y. Computing the indices of Sturm-Liouville eigenvalues for coupled boundary conditions (the EIGENIND-SLP codes)[J]. Journal of Computational and Applied Mathematics, 2008, 220: 490-507.

[6] WANG G X, SUN J. Approximations of eigenvalues of Sturm-Liouville problems in a given region and corresponding eigenfunctions[J]. Pacific Journal of Applied Mathematics, 2011, 3(1/2): 75-96.

[7] 方欣华, 杜涛. 海洋内波基础和中国海内波 [M]. 青岛: 中国海洋大学出版社, 2005.
FANG X H, DU T. Fundamentals of oceanic internal waves and internal waves in the China seas[M]. Qingdao: China Ocean University Press, 2005.

[8] 韩忠杰. 系列连接弹性梁的控制设计与稳定性分析 [D]. 天津: 天津大学, 2007.
HAN Z J. Control design and stability analysis of serially connected elastic beam[D]. Tianjin: Tianjin University, 2007.

[9] 韩忠杰. Timoshenko 梁系统的控制设计与稳定性分析 [D]. 天津: 天津大学, 2010.
HAN Z J. Control design and stability analysis of Timoshenko beam systems[D]. Tianjin: Tianjin University, 2010.

[10] PONTRELLI G, DE MONTE F. Mass diffusion through two-layer porous media: an application to the drug-eluting stent[J]. International Journal of Heat and Mass Transfer, 2007, 50(17/18): 3658-3669.

[11] TITEUX I, YAKUBOV Y. Completeness of root functions for thermal conduction in a strip with piecewise continuous coefficients[J]. Mathematical Models and Methods in Applied Sciences, 1997, 7(7): 1035-1050.

[12] ANDERSSEN R. The effect of discontinuous in density and shear velocity on the asymptotic overtone structure of torsional eigenfrequencies of the earth[J]. Geophysical Journal of the Royal Astronomical Society, 1977, 50(2): 303-309.

[13] FULTON C. Two-point boundary value problems with eigenvalue parameter contained in the boundary conditions[J]. Proceedings of the Royal Society of Edinburgh: Section A: Mathematics, 1977, 77(3-4): 293-308.

[14] GOMILKO A, PIVOVARCHIK V. On basis properties of a part of eigenfunctions of the problem of vibrations of a smooth inhomogeneous string damped at the midpoint[J]. Mathematische Nachrichten, 2002, 245(1): 72-93.

[15] KRUGER R J. Inverse problems for nonabsorbing media with discontinuous material properties[J]. Journal of Mathematical Physics, 1982, 23(3): 396-404.

[16] LAPWOOD E, USAMI T. Free Oscillations of The Earth[J]. Physics and Chemistry of the Earth, 1981, 4(27): 239-250.

[17] OZISIK N. Boundary Value Problems of Heat Conduction[M]. New York: Dove Publications, 1989.

[18] AKCAY O. On the boundary value problem for discontinuous Sturm-Liouville operator[J/OL]. Mediterranean Journal of Mathematics, 2019, 16: 7. https://doi.org/10.1007/s00009-018-1279-5.

[19] AKDOĞAN Z, DEMIRCI M, MUKHTAROV O. Green function of discontinuous boundary-value problem with transmission conditions[J]. Mathematical Methods in the Applied Sciences, 2010, 30(14): 1719-1738.

[20] EVERITT W, ZETTL A. Sturm-Liouville differential operators in direct sum spaces[J]. Rocky Mountain Journal of Mathematics, 1986, 16(3): 497-516.

[21] EVERITT W, ZETTL A. Differential operators generated by a countable number of quasi-differential expressions on the real line[J]. Proceedings of the London Mathematical Society, 1992, 64(3): 524-544.

[22] LI K, SUN J, HAO X L. Eigenvalues of regular fourth-order Sturm-Liouville problems with transmission conditions[J]. Mathematical Methods in the Applied Sciences, 2017, 40(10): 3538-3551.

[23] MUKHTAROV O, YAKUBOV S. Problems for differential equations with transmission conditions[J]. Applicable Analysis, 2002, 81(5): 1033-1064.

[24] MUKHTAROV O, AYDEMIR K. Eigenfunction expansion for Sturm-Liouville problems with transmission conditions at one interior point[J]. Acta Mathematica Scientia(English Series), 2015, 35(3): 639-649.

[25] OLǦAR H, MUHTAROV F. The basis property of the system of weak eigenfunctions of a discontinuous Sturm-Liouville problem[J]. Mediterranean Journal of Mathematics, 2017, 14(3): 114.

[26] SUN J, WANG A, ZETTL A. Two-interval Sturm-Liouville operators in direct sum spaces with inner product multiples[J]. Results in Mathematics, 2007, 50(1): 155-168.

[27] 王桂霞, 孙炯. 一类不连续 Sturm-Liouville 问题特征函数的振动性 [J]. 应用数学学报, 2008, 31(3): 500-513.
WANG G X, SUN J. Oscillatory properties of eigenfunctions of a class of discontinuous Sturm-Liouville problems[J]. Acta Mathematica Applicatae Sinica. 2008, 31(3): 500-513.

[28] HALD O. Discontinuous Inverse Eigenvalue Problems[J]. Communications on Pure and Applied Mathematics, 1984, 37(5): 539-577.

[29] CODDINGTON E. The spectral representation of ordinary self-adjoint differential operators[J]. Annals of Mathematics, 1954, 60(1): 192-211.

[30] NAIMARK M. Linear Differential Operators[M]. New York: Frederick Ungar Publishing Co., 1968.

[31] TITCHMARSH C. Eigenfunction Expansions Associated with Second- Order Differential Equa-

tions (Part Ⅱ) [M]. Oxford: Clarendon Press, 1958.
[32] TITCHMARSH C. Eigenfunction Expansions Associated with Second-Order Differential Equations (Part Ⅰ)[M]. London: Oxford University Press, 1962.
[33] EVERITT W, MÖLLER M, ZETTL A. Discontinuous dependence of the n-th Sturm-Liouville eigenvalue[J]. General Inequalities 7: International Series of Numerical Mathematics, 1997, 123: 145-150.
[34] 曹之江. 高阶极限圆型微分算子的自伴扩张 [J]. 数学学报, 1985, 28(2): 205-217.
CAO Z J. Self-adjoint extensions of higher-order limit circular differential operators[J]. Acta Mathematica Sinica, 1985, 28(2): 205-217.
[35] CAO Z J. On self-adjoint domains of 2-nd order differential operators in limit circle case[J]. Acta Mathematica Sinica, 1985, 1(3): 225-230.
[36] SUN J. On the self-adjoint extensions of symmetric ordinary differential operators with middle deficiency indices[J]. Acta Mathematica Sinica, 1986, 2(2): 152-167.
[37] 王万义. 微分算子的辛结构与一类微分算子的谱分析 [D]. 呼和浩特: 内蒙古大学, 2002.
WANG W Y. The symplectic structure of differential operators and spectra analysis of a class of differential operators[D]. Hohhot: Inner Mongolia University, 2002.
[38] 魏广生, 徐宗本. 奇型 Sturm-Liouville 微分算子的限界自伴扩张 [J]. 数学学报, 2004, 47(2): 305-316.
WEI G S, XU Z B. Bound-limited self-adjoint extensions for singular Sturm-Liouville differential operators[J]. Acta Mathematica Sinica, 2004, 47(2): 305-316.
[39] HAO X L, SUN J, ZETTL A. Canonical forms of self-adjoint boundary conditions for differential operators of order four[J]. Journal of Mathematical Analysis and Applications, 2012, 387(2): 1176-1187.
[40] HAO X L, SUN J, ZETTL A. Fourth order canonical forms of singular self-adjoint boundary conditions[J]. Linear Algebra and its Applications, 2012, 437(3): 899-916.
[41] BAILEY P, GORDON M, SHAMPINE L. Automatic solution of the Sturm-Liouville problem[J]. ACM Transactions on Mathematical Software, 1978, 4: 193-208.
[42] BAILEY P, EVERITT W, ZETTL A. Algorithm 810: The SLEIGN2 Sturm-Liouville code[J]. ACM Transactions on Mathematical Software, 2001, 27(2): 143-192.
[43] GREENBERG L, MARLETTA M. Algorithm 775: The code SLEUTH for solving fourth-order Sturm-Liouville problems[J]. ACM Transactions on Mathematical Software, 1997, 23(4): 453-493.
[44] DAUGE M, HELFFER B. Eigenvalues variation. Ⅰ: Neumann problem for Sturm-Liouville operators[J]. Journal of Differential Equations, 1993, 104(2): 243-262.
[45] KONG Q, ZETTL A. Dependence of eigenvalues of Sturm-Liouville problems on the boundary[J]. Journal of Differential Equations, 1996, 126(2): 389-407.
[46] KONG Q, ZETTL A. Eigenvalues of regular Sturm-Liouville problems[J]. Journal of Differential Equations, 1996, 131(1): 1-19.
[47] KONG Q, WU H, ZETTL A. Dependence of eigenvalues on the problems[J]. Mathematische Nachrichten, 1997, 188(1): 173-201.
[48] KONG Q, WU H, ZETTL A. Dependence of the n-th Sturm-Liouville eigenvalue on the problem[J]. Journal of Differential Equations, 1999, 156(2): 328-354.

References

[49] KONG Q, WU H, ZETTL A. Geometric aspects of Sturm-Liouville problems: I. Structures on spaces of boundary conditions[J]. Proceeding of the Royal Society of Edinburgh, 2000, 130(3): 561-589.

[50] LI K, SUN J, HAO X L. Dependence of eigenvalues of $2n$-th order boundary value transmission problems[J]. Boundary Value Problems, 2017(143). DOI: 10.1186/s13661-017-0876-8.

[51] LI M, AO J J, ZHANG H Y. Dependence of eigenvalues of Sturm-Liouville problems on time scales with eigenparameter-dependent boundary conditions[J]. Open Mathematis, 2022, 20: 1215-1228. DOI: 10.1515/math-2022-0507.

[52] ZHANG M Z, LI K. Dependence of eigenvalues of Sturm-Liouville problems with eigenparameter dependent boundary conditions[J/OL]. Applied Mathematics and Computation, 2020, 378: 125214. https: //doi.org/10.1016/j.amc.2020.125214.

[53] ZHANG M Z, WANG Y C. Dependence of eigenvalues of Sturm-Liouville problems with interface conditions[J]. Applied Mathematics and Computation, 2015, 265: 31-39.

[54] BINDING P, HRYNIV R, LANGER H, et al. Elliptic eigenvalue problems with eigenparameter dependent boundary conditions[J]. Journal of Differential Equations, 2001, 174(1): 30-54.

[55] BINDING P, BROWNE P, WATSON B. Inverse spectral problems for Sturm-Liouville equations with eigenparameter dependent boundary conditions[J]. Journal of the London Mathematical Society, 2000, 62(2): 161-182.

[56] BINDING P, BROWNE P, WATSON B. Transformations between Sturm-Liouville problems with eigenvalue dependent and independent boundary conditions[J]. Bulletin of the London Mathematical Society, 2001, 33(6): 749-757.

[57] BINDING P, BROWNE P, WATSON B. Sturm-Liouville problems with boundary conditions rationally dependent on the eigenparameter: II [J]. Journal of Computational and Applied Mathematics, 2002, 148(1): 147-168.

[58] BINDING P, BROWNE P, WATSON B. Sturm-Liouville problems with boundary conditions rationally dependent on the eigenparameter: I [J]. Proceedings of the Edinburgh Mathematical Society, 2002, 45(3): 631-645.

[59] BROWNE P, SLEEMAN B. A uniqueness theorem for inverse eigenparameter dependent Sturm-Liouville problems[J]. Inverse Problems, 1997, 13(6): 1453-1462.

[60] CHANANE B. Sturm-Liouville problems with parameter dependent potential and boundary conditions[J]. Journal of Computational and Applied Mathematics, 2008, 212(2): 282-290.

[61] DEMIRCI M, AKDOĞAN Z, MUKHTAROV O. Asymptotic behavior of eigenvalues and eigenfunctions of one discontinuous boundary-value problem[J]. International Journal of Computational Cognition, 2004, 2(3): 101-113.

[62] KADAKAL M, MUKHTAROV O. Discontinuous Sturm-Liouville problems containing eigenparameter in the boundary conditions[J]. Acta Mathematica Sinica, 2006, 22(5): 1519-1528.

[63] KERIMOV N, MAMEDOV K. On a boundary value problem with a spectral parameter in the boundary conditions[J]. Sibirskii Matematicheskii Zhurnal, 1999, 40(2): 325-335.

[64] SUN J, WANG W Y. Eigenvalues of a class of regular fourth-order Sturm-Liouville problems[J]. Applied Mathematics and Computation, 2012, 218(19): 9716-9729.

[65] WALTER J. Regular eigenvalue problems with eigenvalue parameter in the boundary conditions[J]. Mathematische Zeitschrift, 1973, 133(4): 301-312.

[66] WANG A P, SUN J, HAO X L, et al. Completeness of eigenfunctions of Sturm-Liouville problems with transmission conditions[J]. Methods and Applications of Analysis, 2009, 16(3): 299-312.

[67] 赵迎春. 内部具有不连续性 Sturm-Liouville 算子的研究 [D]. 呼和浩特: 内蒙古大学, 2018. ZHAO Y C. Sturm-Liouville operators with discontinuity at interior points[D]. Huhhot: Inner Mongolia University, 2018.

[68] WANG A P, SUN J, ZETTL A. Characterization of domains of self-adjoint ordinary differential operators[J]. Journal of Differential Equations, 2009, 246(4): 1600-1622.

[69] ZETTL A. Sturm-Liouville Theory[M]. Providence: American Mathematics Society, 2005.

[70] ZETTL A, SUN J. Self-adjoint ordinary differential operators and their spectrum[J]. Rocky Mountain Journal of Mathematics, 2015, 45(3): 763-886.

[71] GREGUŠ M. Third order linear differential equations: Mathematics and its Applications[M]. Dordrecht: Kluwer Academic Publishers, 1987.

[72] HAO X L, ZHANG M Z, SUN J, et al. Characterization of domains of self-adjoint ordinary differential operators of any order, even or odd[J]. Electronic Journal of Qualitative Theory of Differential Equations, 2017(61): 1-19.

[73] NIU T, HAO X L, SUN J, et al. Canonical forms of self-adjoint boundary conditions for regular differential operators of order three[J]. Operators and Matrices, 2020, 1: 207-220.

[74] HOPKINS B, KOSMATOV N. Third-order boundary value problems with sign-changing solutions[J]. Nonlinear Analysis: Theory, Methods and Applications, 2007, 67(1): 126-137.

[75] LI K, BAI Y L, WANG W Y, et al. Self-adjoint realization of a class of third-order differential operators with eigenparameter dependent boundary conditions[J]. Journal of Applied Analysis and Computation, 2020, 10(6): 2631-2643.

[76] UĞURLU E. Regular third-order boundary value problems[J]. Applied Mathematics and Computation, 2019, 343: 247-257.

[77] UĞURLU E. Third-order boundary value transmission problems[J]. Turkish Journal of Mathematics, 2019, 43(3): 1518-1532.

[78] ALLAHVERDIEV B, BAIRAMOV E, UĞURLU E. Eigenparameter dependent Sturm-Liouville problems in boundary conditions with transmission conditions[J]. Journal of Mathematical Analysis and Applications, 2013, 401(1): 388-396.

[79] AO J J, SUN J, ZHANG M Z. Matrix representations of Sturm–Liouville problems with transmission conditions[J]. Computers and Mathematics with Applications, 2012, 63(8): 1335-1348.

[80] MUKHTAROV O, AYDEMIR K. Minimization principle and generalized Fourier series for discontinuous Sturm-Liouville systems in direct sum spaces[J]. Journal of Applied Analysis and Computation, 2018, 8(5): 1511-1523.

[81] MUKHTAROV O, AYDEMIR K. Two-linked periodic Sturm–Liouville problems with transmission conditions[J]. Mathematical Methods in the Applied Sciences, 2021, 44(18): 14664-14676.

[82] ZHANG M Z. Regular approximation of singular Sturm-Liouville problems with transmission conditions[J]. Applied Mathematics and Computation, 2014, 247: 511-520.

[83] ZHAO Y C, SUN J, ZETTL A. Self-adjoint Sturm-Liouville problems with an infinite number of boundary conditions[J]. Mathematische Nachrichten, 2016, 289(8-9): 1148-1169.

[84] AKCAY O. The representation of the solution of Sturm-Liouville equation with discontinuity conditions[J]. Acta Mathematica Scientia, 2018, 38(4): 1195-1213.

[85] AKDOĞAN Z, DEMIRCI M, MUKHTAROV O. Discontinuous Sturm-Liouville Problems with Eigenparameter- Dependent Boundary and Transmissions Conditions[J]. Acta Applicandae Mathematica, 2005, 86(3): 329-344.

[86] AYDEMIR K. Boundary value problems with eigenvalue-dependent boundary and transmission conditions[J/OL]. Boundary Value Problems, 2014: 131. http: // www.boundary- valueproblems.com/content/2014/1/131.

[87] MU D, SUN J, YAO S. Asymptotic behaviors and Green's function of two-interval Sturm-Liouville problems with transmission conditions[J]. Mathematica Applicata, 2014, 27(3): 658-672.

[88] SHAHRIARI M, AKBARFAM A, TESCHL G. Uniqueness for inverse Sturm-Liouville problems with a finite number of transmission conditions[J]. Journal of Mathematical Analysis and Applications, 2012, 395(1): 19-29.

[89] ZHANG X Y, SUN J. A class of fourth-order differential operator with eigenparameter-dependent boundary and transmission conditions[J]. Mathematica Applicata, 2013, 26(1): 205-219.

[90] PRUESS S, FULTON C. Mathematical software for Sturm-Liouville problems[J]. ACM Transactions on Mathematical Software, 1993, 19(3): 360-376.

[91] BAILEY P, EVERITT W, ZETTL A. The SLEIGN2 Sturm-Liouville code[J]. ACM Transactions on Mathematical Software, 2001, 27(2): 143-192.

[92] DWYER H, ZETTL A. Computing Eigenvalues of Regular Sturm-Liouville Problems[J/OL]. Electronic Journal of Differential Equations, 1994(6). http://ejde.math.swt.edu.

[93] WANG G X, SUN J. A new method of solving the index problem for Sturm-Liouville eigenvalues[J]. Acta Mathematicae Applicatae Sinica(English Series), 2015, 31(4): 1001-1012.

[94] EASTHAM M, KONG Q K, WU H Y, et al. Inequalities among eigenvalues of Sturm-Liouville problems[J]. Journal of Inequalities and Applications, 1999, 3(1): 2-43.

[95] THARWAT M, BHRAWY A, YILDIRIM A. Numerical computation of eigenvalues of discontinuous Sturm-Liouville problems with parameter dependent boundary conditions using sinc method[J]. Numerical Algorithms, 2013, 63(1): 27-48.

[96] 孙炯, 王万义. 微分算子的自共轭域和谱分析-微分算子研究在内蒙古大学三十年 [J]. 内蒙古大学学报 (自然科学版), 2009, 40(4): 469-485.
SUN J, WANG W Y. Characterization of domains of self-adjoint ordinary differential operators and spectral analysis[J]. Journal of Inner Mongolia University, 2009, 40(4): 469-485.

[97] BINDING P, VOLKMER H. Oscillation theory for Sturm-Liouville problems with indefinite coefficients[J]. Proceedings of the Royal Society of Edinburgh, 2001, 131(5): 989-1002.

[98] CAO X F, KONG Q K, WU H Y, et al. Sturm-Liouville problems whose leading coefficient function changes sign[J]. Canadian Journal of Mathematics, 2003, 55(4): 724-749.

[99] FULTON C. On generating theorems and conjectures in spectral theory with computer assistance[M]//HINTON D. Spectral Theory and Computational Methods of Sturm-Liouville problem. CRC Press, 1977: 285-299.

[100] WANG Z, WU H Y. The index problem for eigenvalues for coupled boundary conditions and Fulton's conjecture[J]. Monatshefte für Mathematik, 2009, 157: 177-191.

[101] WANG G X, WANG Z, WU H Y. Relations among eigenvalues of Sturm-Liouville problems with different types of leading coefficient functions[J]. Journal of Mathematical Analysis and

Applications, 2007, 336(2): 1061-1072.

[102] MÖLLER M. On the unboundedness below of the Sturm-Liouville operator[J]. Proceedings of the Royal Society of Edinburgh, 1999, 129(5): 1011-1015.

[103] BINDING P, VOLKMER H. Existence and asymptotics of eigenvalues of indefinite systems of Sturm-Liouville and Dirac type[J]. Journal of Differential Equations, 2001, 172(1): 116-133.

[104] CAO X F, KONG Q K, WU H Y, et al. Geometric aspects of Sturm-Liouville problems: III. Level surfaces of the n-th eigenvalue[J]. Journal of Computational and Applied Mathematics, 2007, 28: 176-193.

[105] PALEY R, WIENER N. Fourier Transforms in the Complex Domain[M]. New York: American Mathematics Society, 1934.

[106] KADETS M. The exact value of the Paley-Wiener constant[J]. Doklady Akademii Nauk SSSR, 1964, 82(6): 1253-1254.

[107] KATZNEL'SON V. Bases of exponential functions in L^2[J]. Funkcjonalna Analiza i Priložen. 1971, 5(1): 37-47.

[108] YOUNG M. An Introduction to Nonharmonic Fourier Series[M]. New York: American Mathematics Society, 1980.

[109] BAI Y L, WANG W Y, WANG G X, et al. Construction and stability of Riesz bases[J/OL]. Journal of Function Spaces, 2018: 5063847. https: //doi.org/10.1155/2018/5063847.

[110] BINDING P, ĆURGUS B. Riesz bases of root vectors of indefinite Sturm-Liouville problems with eigenparameter dependent boundary conditions: I [J]. Operator Theory: Advances and Applications, 2005, 163: 75-95.

[111] FREILING G, YURKO V. Inverse Sturm-Liouville Problem and Their Applications[M]. New York: Nova Science Publishers, 2001.

[112] HARUTYUNYAN T, PAHLEVANYAN A, SRAPIONYAN A. Riesz bases generated by the spectra of Sturm-Liouville problems[J/OL]. Electronic Journal of Differential Equations, 2013, 71. http: //ejde.math.txstate.edu.

[113] HRYNIV R. Uniformly bounded families of Riesz bases of exponentials, sines, and cosines[J]. Mathematical Notes, 2010, 87(3): 510-520.

[114] HE X H, VOLKMER H. Riesz bases of solutions of Sturm-Liouville equations[J]. Journal of Fourier Analysis and Applications, 2001, 7(3): 297-307.

[115] LEVINSON N. Gap and Density Theorems[M]. New York: American Mathematics Society, 1940.

[116] MOISSEV E. On the basis property of systems of sines and cosines[J]. Doklady Akademii Nauk SSSR, 1984, 275(4): 794-798.

[117] OLĞAR H, MUKHTAROV O. Weak eigenfunctions of two-interval Sturm-Liouville problems together with interaction conditions[J]. Journal of Mathematical Physics, 2017, 58(4) : 388-396.

[118] OLĞAR H, MUHTAROV F. The basis property of the system of weak eigenfunctions of a discontinuous Sturm-Liouville problem[J/OL]. Mediterranean Journal of Mathematics, 2017, 14(3). DOI: 10.1007/s00009-017-0915-9.

[119] AYDEMIR K, MUKHTAROV O. Generalized Fourier Series as Green's Function Expansion for Multi-interval Sturm-Liouville Systems[J/OL]. Mediterranean Journal of Mathematics 2017, 14(3). DOI: 10.1007/s00009-017-0901-2.

[120] ALLAHVERDIEV B. Extensions, dilations and functional models of Dirac operators[J]. Integral Equations and Operator Theory, 2005, 51(4): 459-475.

[121] ALLAHVERDIEV B. Non-self-adjoint singular Sturm-Liouville operators in limit-circle case[J]. Taiwanese Journal of Mathematics, 2012, 16(6): 2035-2052.

[122] LI K, SUN J, HAO X L, et al. Spectral analysis for discontinuous non-self-adjoint singular Dirac operators with eigenparameter dependent boundary condition[J]. Journal of Mathematical Analysis and Applications, 2017, 453(1): 304-316.

[123] MUKHTAROV O, AYDEMIR K. New type of Sturm-Liouville problems in associated Hilbert spaces[J/OL]. Journal of Function Spaces, 2014, 2014: 606815. https: //doi.org/10.1155/2014/606815.

[124] WANG Z, WU H Y. Dissipative non-self-adjoint Sturm-Liouville operators and completeness of their eigenfunctions[J]. Journal of Mathematical Analysis and Applications, 2012, 394(1): 1-12.

[125] ATKINSON F. Discrete and Continuous Boundary Value Problems[M]. New York: Academic Press, 1964.

[126] WEIDMANN J. Spectral Theory of Ordinary Differential Operators: Lecture Notes in Mathematics[M]. Berlin: Springer, 1987.

[127] HINTON D. Deficiency indices of odd-order differential operators[J]. Rocky Mountain Journal of Mathematics, 1978, 8(4): 627-640.

[128] KONG Q K, ZETTL A. Linear ordinary differential equations[J]. World Scientific Series in Applicable Analysis: Inequalities and Applications, 1994(3): 381-397.

[129] UĞURLU E, BAIRAMOV E. Spectral analysis of eigenparameter dependent boundary value transmission problems[J]. Journal of Mathematical Analysis and Applications, 2014, 413(1): 482-494.

[130] GOODMAN T, LEE S. Wavelets of multiplicity[J]. Transactions of the American Mathematical Society, 1994, 342(1): 307-324.

[131] LEE C Y, BEARDSLEY R. The generation of long nonlinear internal waves in a weakly stratified shear flow[J]. Journal of Geophysical Research, 1974, 79(3): 453-462

[132] 张善武. 基于变系数 KdV-type 理论模型的南海北部内孤立波传播演变过程研究 [D]. 青岛: 中国海洋大学, 2014.
ZHANG S W. Investigation of the propagation and evolution processes of the internal solitary waves in the northern South China Sea based on the variable coefficients KdV-type theoretical models[D]. Qingdao: Ocean University of China, 2014.

[133] LAMB K, YAN L. The evolution of internal wave undular bores: comparisons of a fully nonlinear numerical model with weakly nonlinear theory [J]. Journal of Physical Oceanography, 1996, 26(12): 2712-2734.

[134] GRIMSHAW R, PELINOVSKY E, TALIPOVA T, et al. Simulation of the transformation of internal solitary waves on oceanic shelves [J]. Journal of Physical Oceanography, 2004, 34(12): 2774-2791.

[135] 蔡树群. 内孤立波数值模式及其在南海区域的应用 [M]. 北京: 海洋出版社, 2015.
CAI S Q. Numerical Model of Internal Solitary Waves and its Application in the South China Sea[M]. Beijing: China Ocean Press, 2015.

[136] FLIEGE M, HUNKINS K. Internal wave dispersion calculated using the Thomson-Haskell method[J]. Journal of Physical Oceanography, 1975, 5(3): 541-548.